THE BSCS
5E
INSTRUCTIONAL MODEL

CREATING
TEACHABLE
MOMENTS

THE BSCS 5E

INSTRUCTIONAL MODEL

CREATING TEACHABLE MOMENTS

RODGER W. BYBEE

National Science Teachers Association

Arlington, Virginia

National Science Teachers Association

Claire Reinburg, Director
Wendy Rubin, Managing Editor
Andrew Cooke, Senior Editor
Amanda O'Brien, Associate Editor
Donna Yudkin, Book Acquisitions Coordinator

Art and Design
Will Thomas Jr., Director
Rashad Muhammad, Cover and interior design

Printing and Production
Catherine Lorrain, Director

National Science Teachers Association
David L. Evans, Executive Director
David Beacom, Publisher

1840 Wilson Blvd., Arlington, VA 22201
www.nsta.org/store
For customer service inquiries, please call 800-277-5300.

NSTA is committed to publishing material that promotes the best in inquiry-based science education. However, conditions of actual use may vary, and the safety procedures and practices described in this book are intended to serve only as a guide. Additional precautionary measures may be required. NSTA and the authors do not warrant or represent that the procedures and practices in this book meet any safety code or standard of federal, state, or local regulations. NSTA and the authors disclaim any liability for personal injury or damage to property arising out of or relating to the use of this book, including any of the recommendations, instructions, or materials contained therein.

Library of Congress Cataloging-in-Publication Data

Bybee, Rodger W.
   The BSCS 5E instructional model : creating teachable moments / by Rodger W. Bybee.
      pages cm
   Includes bibliographical references.
   ISBN 978-1-941316-00-9 (print : alk. paper) -- ISBN 978-1-941316-81-8 (e-book : alk. paper) 1. Science--Study and teaching (Elementary) 2. Science--Study and teaching (Secondary) I. Title. II. Title: Biological sciences curriculum study 5E instructional model.
   Q181.B97 2015
   507.1'2--dc23
                              2015001188

Cataloging-in-Publication Data for the e-book are available from the Library of Congress.

For Nan.
You helped create the 5E Instructional Model and supported its use throughout
your career at BSCS.
This dedication is with my deepest appreciation.

CONTENTS

ENGAGE

Creating Interest in the BSCS 5E Instructional Model

EXPLORE

Perspectives on Instructional Models

EXPLAIN

A Contemporary Discussion

ELABORATE

Expanding and Adapting Your Understanding

EVALUATE

Assessing Understanding and Use

PREFACE

Since the BSCS 5E Instructional Model was developed in the late 1980s, it has been widely implemented in places such as state frameworks and frequently used in articles in professional publications about teaching. This widespread dissemination and use of the model has been, to say the least, amazing. I have often wondered about the extensive application of the model. I have asked questions such as, "What accounts for the model's popularity?" and "Why do teachers embrace the model?" In addition, I have asked whether the BSCS 5E Instructional Model is appropriate for contemporary teaching and learning.

Lest the reader be too surprised, I think the 5E Model's widespread application can be explained by several observations. The first may be the most obvious: The model addresses every teacher's concern—how to be more effective in the classroom. Second, the model has a "common sense" value; it presents a natural process of learning. Finally, the 5 Es are understandable, usable, and manageable by both curriculum developers and classroom teachers.

To my second question about contemporary use, I do believe the BSCS 5E Instructional Model is appropriate for contemporary innovations such as *A Framework for K–12 Science Education*, the *Next Generation Science Standards* (*NGSS*; NGSS Lead States 2013), STEM education, and 21st-century skills.

A Framework for K–12 Science Education, for example, sets forth policies that require integrating three dimensions—science and engineering practices, disciplinary core ideas, and crosscutting concepts. Is it possible to use the 5E Model to meet the challenge of implementing three-dimensional teaching and learning? The *Framework* and *NGSS* require innovations such as constructing explanations, designing solutions, and engaging in argument from evidence. Can practices such as these be addressed within the BSCS model? What about the use of contemporary technologies? Yes, the BSCS 5E Instructional Model can accommodate these contemporary innovations. I used the 5E Model for examples in *Translating the* NGSS *for Classroom Instruction* (Bybee 2013) and will include further discussions later in this book.

I must mention the book's subtitle and theme—creating teachable moments. As a classroom teacher, I experienced times when students were totally engaged. They were caught by phenomena, events, or situations that brought forth a need to know and increased motivation to learn. I am sure most, if not all, classroom teachers have had similar experiences.

PREFACE

When these experiences occur, classroom teachers capture the potential of these teachable moments. Teachers are pleased when this occurs. The common conception of a teachable moment is that it is random and unplanned, that it just occurs from a current event or in the context of a classroom activity, student question, school problem, or other opportunity.

What if you could provide more opportunities for teachable moments? What if teachable moments were not totally random and unplanned, and the probability of an occurrence could be increased through the structuring and sequencing of your lessons? The BSCS 5E Instructional Model described in this book provides classroom teachers with an approach to teaching that changes the emphasis within lessons and provides a sequence that increases the probability of teachable moments.

Here is some context on developing the 5E Model. In the mid-1980s, I assumed the position of associate director of the Biological Sciences Curriculum Study (BSCS). In that position, I helped create the BSCS 5E Instructional Model. At the time, a team of colleagues and I were developing a new program for elementary schools. We needed an instructional model that enhanced student learning and was understood by classroom teachers. Although the instructional model had a basis in learning theory, we avoided the psychological terms and chose to use everyday language to identify the phases of instruction as *engage*, *explore*, *explain*, *elaborate*, and *evaluate*.

When we created the 5E Model, the team and I only had a proposed BSCS program in mind. We had no idea that the instructional model would be widely applied in the decades that followed, commonly modified, and frequently used without reference to or recognition of its origins.

With the experiences of several decades, I made the connection between teachable moments and the BSCS 5E Instructional Model. While I recognized the connection and need for an in-depth discussion of the model, other professional obligations did not allow time to realize the potential in the form of a book. Now, almost three decades later, I have time, and the National Science Teachers Association (NSTA) has given me the opportunity to reflect on the BSCS 5E Instructional Model and consider its origins, history, and contemporary applications.

Before a detailed discussion of this book and the BSCS 5E Instructional Model, a few words of background seem appropriate. In developing the instructional model, we did take several issues into consideration. First, to the degree that it was possible, we wanted to begin with an instructional model that was research based. Hence, we began with the Science Curriculum Improvement Study (SCIS) Learning Cycle because it had substantial evidence supporting the phases and sequence. The additions and modifications we made to the Learning Cycle also had a basis in research.

Second, we realized that the constructivist view of learning requires experiences to challenge students' current conceptions (i.e., misconceptions) and ample time and activities that facilitate the reconstruction of ideas and abilities.

Third, we wanted to provide a perspective for teachers that was grounded in research and had an orientation and purpose for individual lessons. What perspective should teachers have for a particular unit, lesson, or activity? Common terms such as *engage* and *explore* indicated an instructional perspective for teachers. In addition, we wanted to express coherence for lessons within an instructional sequence. How does one lesson contribute to the next, and what was the purpose of the sequence of lessons?

Finally, we tried to describe the model in a manner that would be understandable, usable, memorable, and manageable. All of these considerations contributed to the development of the 5E Instructional Model.

Not surprisingly, I structured this book using the 5E Model. Chapter 1 introduces the engaging theme (I hope) of teachable moments and, very briefly, the BSCS 5E Instructional Model. Chapter 2 explores the historical idea of what can be considered an instructional model. Chapter 3 is an in-depth explanation of the BSCS 5E Instructional Model. Chapter 4 reviews education research supporting instructional models, including the 5Es. Chapters 5, 6, and 7 elaborate on the model's application to *NGSS*, STEM education, 21st-century skills, and implementation in the classroom, respectively. Chapters 8, 9, and 10 present evaluations in the form of questions about the BSCS 5E Model and concluding reflections.

The audience for this book includes curriculum developers, classroom teachers, and those responsible for the professional development of teachers. I have tried to maintain a conversational tone and weave a narrative of education research, the psychology of learning, and the reality of classroom practice.

REFERENCES

Bybee, R. 2013. *Translating the* NGSS *for classroom instruction.* Arlington, VA: NSTA Press.

NGSS Lead States. 2013. *Next Generation Science Standards: For states, by states.* Washington, DC: National Academies Press. *www.nextgenscience.org/next-generation-science-standards.*

ACKNOWLEDGMENTS

I acknowledge and express my gratitude to a team of colleagues that helped create the BSCS 5E Instructional Model. That team included Nancy Landes, Jim Ellis, Janet Carlson Powell, Deborah Muscella, Bill Robertson, Susan Wooley, Steve Cowdrey, and Gail Foster.

The BSCS team who helped prepare *The BSCS 5E Instructional Model: Origins, Effectiveness, and Applications* (Bybee et al. 2006) included Joseph Taylor, April Gardner, Pamela Van Scotter, Janet Carlson Powell, Anne Westbrook, and Nancy Landes. In addition, other BSCS staff contributed by assisting with the research: Samuel Spiegel, Molly McGarrigle Stuhlsatz, Amy Ellis, Barbara Resch, Heather Thomas, Mark Bloom, Renee Moran, Steve Getty, and Nicole Knapp.

When I began writing this book, I contacted Pamela Van Scotter, then acting director of BSCS. After telling her of my intention and asking about the use of BSCS reports and materials, she immediately and unconditionally gave permission. I thank her and acknowledge her long and deep support of BSCS and my work.

BSCS staff provided support for this work. I especially acknowledge Joe Taylor for providing articles and information on research supporting the 5E Model. Stacey Luce, the production coordinator, was most helpful with permissions and art for this book. I appreciate her contribution.

Appreciation must be expressed for my editors at NSTA Press. Claire Reinburg has consistently supported the publication of my works, and Wendy Rubin has improved on every draft manuscript. And just for Wendy—go Rockies! Maybe next year.

A special acknowledgment goes to Linda Froschauer, editor of *Science and Children*. Late in 2013, Linda asked me to prepare a guest editorial on the BSCS 5E Instructional Model. Linda's request and the preparation of the editorial initiated the long-overdue work on this book.

Reviewers for this manuscript included Pamela Van Scotter, Harold Pratt, Nancy Landes, Karen Ansberry, and Nicole Jacquay.

I express my sincere and deep gratitude to Nancy Landes. Nancy was on the BSCS team that created the BSCS 5E Instructional Model, incorporated the model in numerous BSCS programs she developed, and completed a thorough review of an early draft of this book. As an expression of my appreciation, I have dedicated this book to Nancy.

ACKNOWLEDGMENTS

Byllee Simon has been my assistant for five years. She and I have worked closely on five books. My debt to Byllee is broad and deep.

Finally, Kathryn Bess read the manuscript. Her comments continually reminded me that teachers will use the 5E Model, and her recommendations brought sensitivity and a personal touch to the book. Kathryn has long supported my work. I am indebted to her and extend my appreciation and gratitude.

Rodger W. Bybee
Golden, Colorado
October 2014

REFERENCE

Bybee, R. W., J. A. Taylor, A. Gardner, P. Van Scotter, J. C. Powell, A. Westbrook, and N. Landes. 2006. *The BSCS 5E Instructional Model: Origins, effectiveness, and applications.* Colorado Springs, CO: Biological Sciences Curriculum Study (BSCS).

ABOUT THE AUTHOR

Rodger W. Bybee, PhD, was most recently the executive director of the Biological Sciences Curriculum Study (BSCS), a nonprofit organization that develops curriculum materials, provides professional development, and conducts research and evaluation for the education community. He retired from BSCS in 2007.

Prior to joining BSCS, Dr. Bybee was executive director of the National Research Council's (NRC) Center for Science, Mathematics, and Engineering Education (CSMEE), in Washington, D.C. From 1986 to 1995, he was associate director of BSCS, where he was principal investigator for four new National Science Foundation (NSF) programs: an elementary school program called *Science for Life and Living: Integrating Science, Technology, and Health*; a middle school program called *Middle School Science & Technology*; a high school program called *Biological Science: A Human Approach*; and a college program called *Biological Perspectives*. He also served as principal investigator for programs to develop curriculum frameworks for teaching about the history and nature of science and technology for biology education at high schools, community colleges, and four-year colleges, as well as curriculum reform based on national standards.

Dr. Bybee participated in the development of the *National Science Education Standards*, and from 1993 to 1995 he chaired the content working group of that NRC project. From 1990 to 1992, Dr. Bybee chaired the curriculum and instruction study panel for the National Center for Improving Science Education (NCISE). From 1972 to 1985, he was professor of education at Carleton College in Northfield, Minnesota. He has been active in education for more than 40 years and has taught at the elementary through college levels.

Dr. Bybee received his BA and MA from the University of Northern Colorado and his PhD from New York University. Dr. Bybee has written about topics in both education and psychology. He has received awards as a Leader of American Education and an Outstanding Educator in America, and in 1979 he was named Outstanding Science Educator of the Year. In 1989, he was recognized as one of 100 outstanding alumni in the history of the University of Northern Colorado. In April 1998, the National Science Teachers Association (NSTA) presented Dr. Bybee with NSTA's Distinguished Service to Science Education Award. Dr. Bybee chaired the Science Forum and Science Expert Group (2006) for the Programme for International Student Assessment of the OECD (PISA). In 2007, he received the Robert H. Carleton Award, NSTA's highest honor for national leadership in the field of science education.

Although he has retired from BSCS, Dr. Bybee continues to work as a consultant.

ENGAGE

Creating Interest in the BSCS 5E Instructional Model

What Are Teachable Moments, and How Are They Created?

Let's begin with several questions. I ask that you take a moment or two and reflect on my questions and your answers. What is your description of a teachable moment? Have you ever experienced a time when students were highly motivated to learn? How would you describe that situation? What did you do? What did the students do? If you wanted to create another teachable moment for students, what would you say or do? Here are other questions:

What is your *primary* frame of reference for teaching?

   a. lesson (1 or 2 days)

   b. unit (2–4 weeks)

   c. semester (14–16 weeks)

   d. all of the above

What is your typical sequence of teaching?

   a. Present information, give examples, practice, test.

   b. Describe context, inform students, verify student learning, test.

   c. Ask a question, introduce ideas, have students apply ideas, test.

   d. Other (Please describe.)

Are you open to thinking differently about your teaching?

   a. yes

   b. no

   c. maybe

Most teachers' primary frame of reference for teaching is the daily lesson. This view is followed by the unit. This process is not unusual or bad. As a professional, you begin each day's work with the lesson. The typical sequence of teaching is some variation of "Present ideas, give examples, students practice or apply, test." Granted, there are variations based on content, difficulty, students' interest, and motivation.

Would you consider a different approach to teaching? I hope your response is yes or maybe. Assuming this is the case, we will continue.

This chapter introduces two themes of the book. First, there is a discussion of teachable moments. Second, I summarize the BSCS 5E Instructional Model, which will help you create teachable moments and use them as the foundation for student learning. The instructional model is a different way of thinking about teaching, and it includes lessons and expands one's perspective from lessons to an instructional sequence.

TEACHABLE MOMENTS

Like classroom teachers at all levels and disciplines, you have probably experienced teachable moments. Teachable moments are those positive distractions from planned lessons where students are engaged and a teacher has the opportunity to explore ideas and provide an explanation or insight. These are exciting, even magical, moments for teachers. Let's look more closely at the idea of teachable moments.

What Is a Teachable Moment?

Most teachers know when they see them in students, but what are teachable moments? In education, a teachable moment is generally perceived as a time when students are motivated to learn. Discussions of teachable moments use terms such as *unplanned educational opportunity*, which is an unanticipated or unscheduled time when the probability of learning is greatly enhanced.

The important point is that at that moment, a student is engaged and eager to learn, and a teacher can easily teach a particular idea or skill. By the way, we all have teachable moments. There are times and situations where all of us want to learn—and do.

When Does a Teachable Moment Occur?

The short answer: anytime. This is why common definitions use words such as *unplanned* and *unscheduled*. That said, from a student's perspective, a key factor in teachable moments is that they occur when an experience has significant personal meaning—that is, the event is important, has consequences, or causes puzzlement for the student. This, then, is a key difference between the content of state and local curriculum frameworks, many lessons, and students' interests and motivation to learn.

Why Does a Teachable Moment Occur?

Teachable moments occur when individuals experience something they recognize and that has meaning, but they cannot formulate an explanation for the phenomenon or experience. The experience is within their cognitive grasp but beyond their full understanding. The verbal evidence of this situation is use of words that express puzzlement, questions, or curiosity, such as *why*, *what*, and *how*, followed by the individual's expression of the phenomenon, event, or situation. An example might be a child's response to an earthquake and the subsequent question, "What causes earthquakes?" Children are especially interested in understanding the world around them and ask questions that present teachable moments.

At a slightly deeper level, the student is expressing cognitive disequilibrium with phenomena in the classroom, school, or environment. In short, the student's current knowledge and understanding do not provide an explanation for something he or she has experienced.

How Can You Create Teachable Moments?

As a classroom teacher, you do not have to wait for something out of the blue; you can create teachable moments by using a sequence of lessons that includes engaging experiences and activities for students, but the experiences should be beyond students' immediate grasp. Imagine using an instructional sequence that begins with an experience of high interest but is beyond students' understanding, and then the lessons provide opportunities for students to sort out their ideas and try to explain the initial situation as the sequence continues.

This leads you to the moment to help students gain knowledge and understanding of the experience. Then, you provide a situation where students have to apply their new knowledge to another situation. Finally, students and the teacher conclude with an assessment. Figure 1.1 shows how this process can work.

I have just used general terms to describe the BSCS 5E Instructional Model, an approach to teaching that centers on important content and abilities and that increases the opportunities for teachable moments.

In the next section, I describe the BSCS 5E Instructional Model in detail so you will have a context for the background, connections, and implications I make in subsequent chapters on the history of instructional models, education research, the *Next Generation Science Standards* (*NGSS*), STEM education, and 21st-century skills.

Next, I present an introduction to the BSCS 5E Instructional Model.

Figure 1.1. The Process of a Teachable Moment

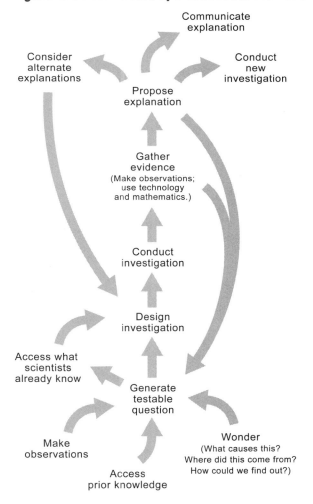

Copyright © 2012 BSCS. Image from BSCS. 2012. *BSCS Middle School Science. www.elearn.bscs.org*

THE BSCS 5E INSTRUCTIONAL MODEL

The discussion here is based on a guest editorial I prepared for *Science and Children* (Bybee 2014). The BSCS 5E Instructional Model consists of five phases of learning: *engage, explore, explain, elaborate,* and *evaluate*. This section describes each phase of the instructional model, and Figure 1.2 shows a sample 5E Model process. Table 1.1 (p. 8) shows connections to teachable moments for each phase of the 5E Model.

Figure 1.2. An Example of the BSCS 5E Instructional Model

Chapter 2
Earth's Heat Engine

Evaluate
Hurricanes are important factors in ocean heat transport.

Engage
Scientists analyze data to predict hurricanes and their impact.

Heat moves around Earth by various natural processes

Elaborate
Volcanoes facilitate heat movement from inside Earth.

Explore
Matter can transport heat.

Explore/Explain
Temperature, salinity, and the rotation of Earth affect the circulation of ocean water and heat transport.

Explain
The atmosphere is important for heat transport around the Earth.

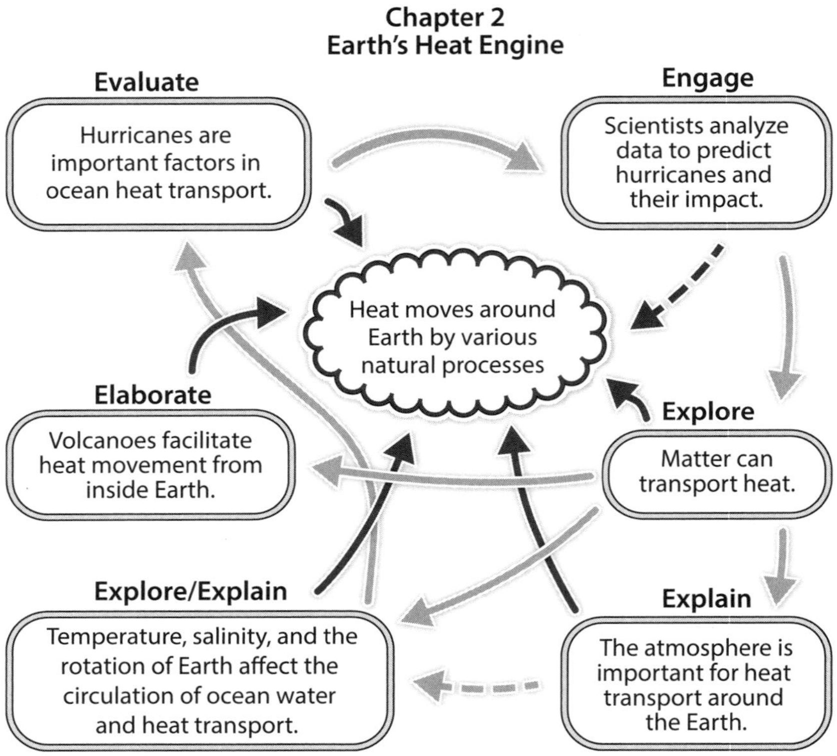

Copyright © 2014 BSCS.

Engage

The goal of this phase is to capture the students' attention. Get the students focused on a situation, event, demonstration, or problem that involves the content and abilities that are the goals of your instruction. From a teaching point of view, you might ask a question, pose a problem, or present a discrepant event as a strategy to engage learners. From a student's perspective, engagement may be the origin of a teachable moment if the student looks puzzled or says, "How did that happen?" or "I have wondered about that." You have engaged the student mentally and initiated a teachable moment. Students may express some ideas, but the ideas are not clear and may not be conceptually accurate.

Students engaged in an activity.
Copyright © 2012 BSCS. Image from BSCS. 2012. *BSCS Middle School Science. www.elearn.bscs.org*

The engagement need not be a full lesson, but often it is. It might be as brief as a question, short demonstration, or presentation of a current event. You might, for example, provide a brief description of a phenomena and ask how the students would explain the situation. The main point is that the experience has a meaningful context and students are thinking about content related to your goals for the instructional sequence. Students do not have an explanation for the experience. Words that describe students' responses include *mystified, perplexed,* or *puzzled*. Although students may be puzzled, their engagement suggests they realize the puzzle can be solved.

The *engage* phase also gives you an opportunity to find out where students are in their thinking about the content or phenomenon. What is their current understanding of the content? Do they have clear misconceptions? Are their misconceptions part of their puzzlement? Noting student misconceptions does not detract from the teachable moment. In fact, it can enhance your ability to continue to engage students with ideas that help them think about the adequacy or inadequacy of their current ideas when contrasted with an explanation for the phenomena.

Explore

In the *explore* phase, students participate in activities that provide the time and opportunities to resolve the mental disequilibrium or dissonance of the engagement experience. Students explore their understanding of the teachable moment's content. The lesson or lessons provide concrete experiences through which students wrestle with their current conceptions and demonstrate their abilities as they try to resolve the puzzlements of the *engage* phase.

Exploration experiences should be designed for subsequent introduction and description of the concepts and skills of the instructional sequence. Students should have time and opportunities to formulate their explanations, investigate phenomena, discuss ideas, and develop their cognitive and physical abilities before they layer on formal language of a discipline, such as a scientific explanation.

The teacher's role in the *explore* phase is to initiate the activity, describe appropriate background, and provide adequate materials and experiences to help resolve students' disequilibrium. After this, the teacher steps back and becomes a facilitator and coach, listening, observing, and guiding students to clarify their understanding as they begin reconstructing their understanding.

Explain

Keeping students connected to, and explaining, the teachable moment is the emphasis of this phase. The concepts, practices, and abilities with which students were engaged and subsequently explored are now made clear and comprehensible. The teacher directs students' attention to key aspects of the prior phases and first asks students for their explanations.

An important aspect of this phase is to begin by having students express their explanations. After students reveal their explanations during the initial phase, the teacher introduces concepts directly and explicitly. Using *NGSS*, for example, the disciplinary core ideas—including vocabulary, science and engineering practices, and crosscutting concepts—are presented clearly and simply. Prior experiences from the *engage* and *explore* phases should be used for context of the explanation.

Teachers commonly use verbal explanations, but use of video, the web, text, software, or other technologies also may provide explanations. This phase is a teacher's common response to a teachable moment.

Elaborate

In this phase, the students are involved in learning experiences that expand and enrich the concepts and abilities developed in the prior phases. The intention is to facilitate the transfer of concepts and abilities to closely related but new situations. A key point for this phase: Use activities that are challenging but achievable by the students.

Students explaining their ideas to one another.
Copyright © 2012 BSCS. Image from BSCS. 2012. *Technology in practice: Applications and innovations. www.kendallhunt.com/technologyinpractice*

The elaboration process involves interaction among students, between the teacher and students, and with other sources, such as written material, databases, simulations, and web-based searches.

This phase extends the teachable moment and emphasizes the transfer of learning from the original context to new but related situations.

Evaluate

Finally, teachers and students should receive feedback on the adequacy of their explanations and abilities. Clearly, informal evaluations will occur beginning with the initial phase of the instructional sequence. But as a practical matter, teachers must assess and report on education outcomes. Hence, the *evaluate* phase.

The *evaluate* phase should involve students in activities that are consistent with those of prior phases and designed to assess their explanations. As part of the evaluate phase, the teacher should determine in advance the evidence for student learning and a means of obtaining the evidence.

Table 1.1. The BSCS 5E Instructional Model and Teachable Moments

PHASES OF THE INSTRUCTIONAL MODEL	CONNECTIONS TO TEACHABLE MOMENTS
Engagement: This phase of the instructional model initiates the learning task. The activity should make connections between past and present learning experiences, anticipate future activities, and focus students' thinking on the learning outcomes. This phase also allows teachers to discover current conceptions (i.e., misconceptions) the students may hold. In this way, the teacher may help students confront those misconceptions in upcoming experiences, thus creating additional teachable moments.	Students identify meaningful events and become mentally engaged by experiences that generate questions or problems. Cognitive conflict, dissonance, or disequilibrium may be expressed by students through comments or questions—such as, "I wondered about that" or "How did that happen?"
Exploration: This phase of the instructional model provides students with a common base of experiences within which they identify and develop current concepts, processes, and skills.	Students try to answer their engaging questions and express their current knowledge and skills. While pursuing resolution for the teachable moment, they make observations, gather information, discuss problems with peers, and consider possible explanations. In this phase, students begin to resolve their cognitive disequilibrium.
Explanation: This phase focuses students' attention on a particular aspect of their understanding and provides opportunities for them to express their conceptual understanding or demonstrate their skills. This phase also provides opportunities for teachers to introduce a formal label or definition for a concept, process, or skill.	Students propose explanations—answers to questions or solutions to problems that created the teachable moment. Then, the teacher explains and clarifies the concepts, skills, or abilities. The teacher's explanation should be brief, direct, and as simple as possible. Students experience a new equilibrium.
Elaboration: This phase challenges and extends students' conceptual understanding and allows further opportunity for students to practice desired skills and behaviors. Through new experiences, the students develop deeper and broader understanding, more information, and better skills.	Students are confronted with a related but new activity that requires the transfer of the concepts, skills, or abilities. Here, a new teachable moment may occur, but it should be minor and connect to the explanation developed in the prior phases.
Evaluation: This phase has students assess their understanding and abilities and provides opportunities for teachers to evaluate student progress toward the learning outcomes.	Students express their knowledge and abilities in an assessment. The assessment should reveal the adequacy of their understanding. The context for the assessment should be a new but related activity that clearly requires the knowledge and abilities developed in the instructional sequence.

CONCLUSION

This chapter serves as an introduction to the primary themes of teachable moments and an instructional model that will help you create moments when students will be excited to learn.

With this brief introduction, one can see the rich opportunities of the BSCS 5E Instructional Model. This model will help teachers bridge the gap between research on learning and the realities of classrooms. In addition, once a teacher understands the aims, orientation, and flexibility of the five phases, he or she can incorporate the unique demands of the *Common Core State Standards*, *NGSS*, and other state and local standards.

The chapters that follow provide more background, details, and examples that will help you create teachable moments in your classroom. These chapters present the history of instructional models and make connections to the *NGSS*, STEM education, and other contemporary topics that will help you implement the BSCS 5E Instructional Model.

REFERENCE

Bybee, R. 2014. The BSCS 5E Instructional Model: Personal reflections and contemporary implications. *Science and Children* 51 (8): 10–13.

EXPLORE

Perspectives on Instructional Models

Exploring Historical Examples of Instructional Models

Did anyone ever propose an instructional sequence similar to the BSCS 5E Instructional Model? The short answer is yes. The BSCS 5E Model has an orientation similar to others in education history. This chapter discusses several historical predecessors to the contemporary 5E Model (Bybee 1997).

JOHANN FRIEDRICH HERBART

Johann Friedrich Herbart, a German philosopher, influenced American education thought in the late 1800s and early 1900s. For Herbart, the primary purpose of education was development of character, and the process of developing character began with cultivating the student's interest. Herbart considered concepts to be the fundamental building blocks of the mind and, thus, justification for including a concept in a course of study. In a contemporary sense, Herbart was interested in the creation and development of knowledge and cognitive abilities that would contribute to an individual's character. Herbart's philosophy contrasted with another model that proposed that the purpose of education was to exercise the mind.

Foundations for Teaching

Herbart described two ideas as foundations for teaching: interest and conceptual understanding. The first principle of effective instruction consisted of the student's interest in the subject. This provides a connection to personal meaning and contexts relative to teachable moments. Herbart suggested two types of interest: one based on direct experiences with the natural world and the second based on social interactions. Instruction can quite easily use the natural world and capitalize on the curiosity of students. In addition, teachers can introduce objects from the natural world and use them to help students accumulate a rich set of sense impressions. Herbart suggested the observation and collection of living organisms and the introduction of tools and machines. As teachers introduce lessons, they should take into account and make connections to prior experiences (Herbart 1901).

Herbart's model also implies that teachers recognize the social interests of students and interactions with other individuals. Thus, teaching should incorporate opportunities for social interaction among students and between students and the teacher.

The second principle of Herbart's model resides in the association of sense perceptions with generalizations or principles, or the formation of concepts. For Herbart, sense perceptions of objects, organisms, and events were essential, but they were not sufficient in and of themselves for the development of the mind. An important theme in Herbart's model is the coherence of ideas. That is, each new idea must be related to ideas students currently believe. Said in contemporary terms, a student's prior knowledge and current conceptions should be recognized as a part of instruction.

Herbart's Instructional Model

Herbart's ideas can be synthesized into an instructional model (see, for example, Compayre 1907; DeBoer 1991). Teachers would begin with the current knowledge and experiences of the student and present ideas that easily related to those concepts. Second, the teacher introduces new ideas that connect with current knowledge and help students slowly construct more elaborate concepts. According to Herbart (1901), the best pedagogy allowed students to discover the relationships among different, but related, experiences. The teacher would guide, question, and suggest connections and relationships through indirect methods and experiences. The third step in the inferred instructional model involves formal instruction, where the teacher systematically explains ideas that the student could not be expected to discover independently. In the final step of Herbart's model, teachers ask students to demonstrate their conceptual understanding by applying the concepts to new situations. Students would solve problems, write essays, and perform tasks that demonstrate their understanding of the concepts. Herbart's model is one of the first systematic approaches to teaching and has been used in various forms by educators for more than 100 years (DeBoer 1991). Table 2.1 summarizes Herbart's instructional ideas.

Table 2.1. A Synthesis of Herbart's Instructional Model

PHASE	INSTRUCTIONAL STRATEGY
Preparation	The teacher uses techniques that identify students' prior experiences and current concepts.
Presentation	The teacher introduces new experiences and makes connections to prior experiences.
Generalization	The teacher directly explains ideas and develops concepts for the students.
Application	The teacher provides experiences where the students demonstrate their understanding by applying concepts in new contexts.

JOHN DEWEY

John Dewey began his career as a teacher. Dewey's early experiences influenced the connection between Dewey's conception of thinking and processes associated with science and technology. In *How We Think* (Dewey 1933 [1910]), Dewey outlined what he called a complete act of thought and described what he proposed were indispensable traits of reflective thinking. The traits included (1) defining the problem, (2) noting conditions associated with the problem, (3) formulating a hypothesis for solving the problem, (4) elaborating the value of various solutions, and (5) testing the ideas to see which provided the best solution for the problem.

Foundations for Teaching

In *Democracy and Education*, Dewey (1916) further described the relationship between experience and thinking. He summarized the general features of the reflective experience as

> *(i) perplexity, confusion, doubt, due to the fact that one is implicated in an incomplete situation whose full character is not yet determined; (ii) a conjectural anticipation—a tentative interpretation of the given elements, attributing to them a tendency to affect certain consequences; (iii) a careful survey (examination, inspection, exploration, analysis) of all attainable consideration which will define and clarify the problem in hand; (iv) a consequent elaboration of the tentative hypothesis to make it more precise and more consistent, because squaring with a wider range of facts; (v) taking one stand upon the project hypothesis as a plan of action which is applied to the existing state of affairs; doing something overtly to bring about the anticipated results, thereby testing the hypothesis. (p. 150)*

Dewey suggests an instructional approach based on experience that engages reflective thinking by students. In contemporary terms, doing hands-on activities is important but inadequate; those experiences also must be minds-on. Dewey suggests that a worthwhile instructional sequence must provide students with the opportunities to formulate and test hypotheses and thus engage in the reflective thinking process.

In a later volume, *Experiences and Education* (1938), Dewey makes the case for experiences as the basis for many dimensions of learning. He argues that educational experiences consist of two features: They have continuity, and they include interactions with others. Dewey notes that the principle of continuity "means, nevertheless that the future has to be taken into account at every stage of the educational process" (Dewey 1938, p. 47) and "the principle of interaction makes it clear that failure of adaptation of material to needs and capacities of individuals may cause an experience to be non-educative" (Dewey 1938, pp. 46–47). Later, he states:

> *It is a cardinal precept of the newer school of education that the beginning of instruction shall be made with the experience learners already have; that this experience and the capacities that have been developed during its course provide the starting point for all further learning. (Dewey 1938, p. 74)*

Dewey concludes, "I am not so sure that the other condition, that of orderly development toward expansion and organization of subject-matter through growth of experience, receives as much attention" (Dewey 1938, p. 74).

Dewey's Instructional Model

Underlying these general recommendations, one can detect a model of instruction. Table 2.2 summarizes Dewey's implied instructional model. I synthesized this model from Dewey's statements in *Democracy and Education* (1916), *Experiences and Education* (1938), *How We Think* (Dewey 1933 [1910]), and the Commission on Secondary School Curriculum (1937) report *Science in General Education*.

Table 2.2. Synthesis of Dewey's Instructional Model

PHASE	INSTRUCTIONAL STRATEGY
Sensing perplexing situations	The teacher presents an experience where the students sense a problem.
Clarifying the problem	The teacher helps the students identify and formulate the problem.
Formulating a tentative hypothesis	The teacher provides opportunities for students to form possible solutions (i.e., hypotheses) and establish a relationship between the perplexing situation and previous experiences.
Testing hypothesis	The teacher allows students to try various means, including imaginary, pencil-and-paper, and concrete investigations to test the hypothesis.
Revising rigorous tests	The teacher suggests tests that result in acceptance or rejection of the hypothesis.
Acting on the solution	The teacher asks the students to develop a statement that communicates their conclusions and expresses possible actions.

A Unit Method of Instruction

In the same period as John Dewey's writing, I noted an article by R. S. Howard in which he described a "Unit Method of Instruction" (Howard 1927). The article shows a remarkable alignment of terms and ideas used in later instructional models. Here is the sequence proposed by Howard: *exploration, presentation* (preceded by reading preparation), *assimilation, organization, recitation,* and *examination.*

In 1950, a variation of John Dewey's instructional model emerged in science methods textbooks (Heiss, Obourn, and Hoffman 1950). The authors based their "learning cycle" (their term) on Dewey's complete act of thought. Table 2.3 presents that learning.

Table 2.3. Heiss, Obourn, and Hoffman's Learning Cycle

PHASE	SUMMARY
Exploring the unit	Students observe demonstrations to raise questions, propose a hypothesis to answer questions, and plan for testing.
Experience getting	Students test the hypothesis, collect and interpret data, and form a conclusion.
Organization of learning	Students prepare outlines, results, and summaries; they take tests.
Application of learning	Students apply information, concepts, and skills to new situations.

ROBERT KARPLUS

Robert Karplus was a physicist at the University of California, Berkeley. He earned his PhD in chemical physics from Harvard when he was 21 years old. As a physicist, Karplus was both a theoretician and experimentalist, which gave him scientific perspectives that likely contributed to his formulation of a learning cycle for a school science program.

Foundations for Teaching

In the late 1950s, Karplus became concerned that children in his daughter's elementary school were being taught science by reading textbooks. So he visited his daughter's class and conducted science demonstrations for the children. The children were quite intrigued, and even excited, by the teachable moments created by the demonstrations, but they did not seem to learn any science. Karplus realized he needed to know more about how children learn to be an effective teacher (Fuller 2002; Stage 2006).

Karplus discovered the work of Jean Piaget and visited Geneva, Switzerland, to study with Piaget in spring 1961. Children's development from concrete to abstract reasoning and the process of self-regulation (i.e., the process of equilibration) influenced the way Karplus approached education research, curriculum development, and formulation of an instructional model referred to as the Learning Cycle.

The Learning Cycle

In the early 1960s, J. Myron Atkin and Robert Karplus (1962) proposed a systematic approach to instruction. The Learning Cycle they proposed had three phases—*exploration, invention,* and *discovery* (see Table 2.4, p. 19). This instructional model became a foundational aspect of the Science Curriculum Improvement Study (SCIS), a project led by Robert Karplus and Herbert Thier.

The SCIS Learning Cycle was influenced by other individuals and curriculum projects of the era. For example, the classic article "Messing About in Science" by David Hawkins (1965) describes a teaching model using a circle, triangle, and square as symbols. In general, the symbols represented phases of an instructional model similar to the Learning Cycle. That model included unstructured exploration and multiple programmed experiences and didactic instruction.

In past decades, the Learning Cycle has undergone elaboration, modification, and application to different education settings. In addition, an analysis of elementary programs indicated that SCIS was one of the most effective programs (Shymansky, Kyle, and Alport 1983). These positive effects on learning relate at least in part to the Learning Cycle. The Learning Cycle has also been central to a proposed theory of instruction (Lawson, Abraham, and Renner 1989).

Using the Learning Cycle creates situations in which the processes based on Piaget's ideas of assimilation, accommodation, and organization occur (Renner and Lawson 1973; Lawson, Abraham, and Renner 1989). Here is a description of the three phases—*exploration, invention, and discovery*—with elaboration based on Piagetian psychology:

- *Exploration* refers to the relatively unstructured experiences (i.e., teachable moments) in which students gather information. In a Piagetian perspective, this phase involves disequilibrium and, predominantly, the process of assimilation.

- *Invention* refers to a formal statement (often the definition) of a new concept. Following the exploration, the invention phase begins the process of accommodation that allows interpretation of newly acquired information through the restructuring of prior concepts.

- The *discovery* phase involves application of the new concept to another novel situation. During this phase, the learner continues to move closer to a state of equilibrium and a new level of cognitive organization (integration of the new concept with related concepts).

A number of studies have shown that the Learning Cycle has many advantages when compared with other approaches to instruction, specifically the transmission model of teaching. These studies are summarized in Abraham and Renner (1986). Jack Renner and his colleagues (Renner, Abraham, and Birnie 1985; Abraham and Renner 1986; Renner, Abraham, and Birnie 1988) have investigated, respectively, the form of acquisition of information in the Learning Cycle, the sequencing of phases in the Learning Cycle, and the necessity of all phases of the Learning Cycle. These studies have generally supported use of the Learning Cycle as originally proposed by Atkin and Karplus. Research on discovery, guided discovery, and statement-of-rule learning (Egan and Greeno 1973; Gagne and Brown 1961; Roughead and Scandura 1968) supports the "sequencing and necessity" requirements drawn by Renner and his colleagues.

Although the form and structure of the Atkin and Karplus Learning Cycle (1962) have undergone little revision, researchers have offered different interpretations of each phase of the cycle. Abraham and Renner (1986) refer to the exploration phase as *gathering data*. This interpretation provided more structure for the assimilation of new material and restricts Atkin and Karplus's original notion that exploration should provide students with common experiences, regardless of whether those experiences involve gathering data in a laboratory sense.

Lawson (1988) describes the invention phase as *concept introduction*, suggesting that because the new concept is not fully developed during this phase, the learner does not truly invent the concept. Lawson further suggests that the appropriate label for this phase may be *term introduction* because only the vocabulary associated with the new concept is learned at this point.

Renner renamed the discovery phase *expansion*, taken from the idea that the learner actually expands on the new concept and is involved in the Piagetian process of organization. Lawson (1988) suggested a more restrictive interpretation of this phase when he used the term *concept application*. One should be aware, however, that in applying the concept to new situations, the learner may still be in the process of restructuring or reconstructing the concept. Lawson (1995) has provided an excellent detailed history of the development and modifications of the Learning Cycle.

Those interested in detailed discussion of Robert Karplus, his education research, and the Learning Cycle should refer to *A Love of Discovery* (Fuller 2002), *A Theory of Instruction* (Lawson, Abraham, and Renner 1989), and *The Learning Cycle* (Marek and Cavallo 1997).

Table 2.4. Atkin and Karplus Learning Cycle

PHASE	INSTRUCTIONAL STRATEGY
Exploration	Teacher or curriculum provides initial experience with phenomena.
Invention	The teacher introduces new terms associated with concepts that are the object of study.
Discovery	The teacher provides experiences for students to apply concepts and use of terms in related, but new, situations.

THE BSCS 5E INSTRUCTIONAL MODEL

This section describes development of the BSCS 5E Instructional Model. I take the liberty to make this more personal as I was directly involved with the formulation of the model. While I led the team, I fully acknowledge the contributions made by coworkers, especially

Nancy Landes, Jim Ellis, Janet Carlson, Deborah Muscella, Bill Robertson, Susan Wooley, Steve Cowdrey, Terry Spencer, and Gail Foster.

At this point, I must also note the contributions of Roger Johnson. At a time after the BSCS team had a general formulation of the instructional model, I had breakfast with Roger. After I described the model using terms for the Learning Cycle, Roger suggested using descriptive words with the same initial letter, as that would help teachers and other educators recall and use the model. After some thought, the BSCS 5E Instructional Model was born.

Foundations for Teaching

In the mid-1980s, BSCS received a grant from IBM to conduct a study that would produce design specifications for a new science and health curriculum for elementary schools. The results of that study were published as *New Designs for Elementary School Science and Health* (BSCS and IBM 1989). Among the innovations that resulted from this design study was the BSCS 5E Instructional Model. As mentioned earlier, the BSCS model has five phases: *engagement, exploration, explanation, elaboration,* and *evaluation.* When formulating the BSCS 5E Instructional Model, I consciously began with the Learning Cycle. The psychology underlying the BSCS 5E Instructional Model also was consistent with the model implemented earlier by Karplus and Thier, in that it was primarily a Piagetian orientation. I was familiar with the work of Jean Piaget, as I completed a book titled *Piaget for Educators* (Bybee and Sund 1982). Furthermore, I had been thinking about and considering a modification of the Learning Cycle and used it in the arrangement of chapters and the structure of individual chapters. In a note to readers, I described my use of a modified Learning Cycle: "The book itself and most of the chapters are divided into four sections: Exploration, Explanation, Extension, and Evaluation" (Bybee and Sund 1982, p. xiii).

As a graduate student in the late 1960s, I had studied Piaget's theory and its application to teaching. I was especially interested in the process of equilibration as the means for changing an individual's cognitive structures. Also in the late 1960s, I was invited to Lawrence Hall of Science and spent a week with Robert Karplus, Herb Thier, Chet Lawson, and their colleagues. In this period, I also taught elementary school science and used units from the SCIS. This discussion provides some background and context for the Learning Cycle and the subsequent modifications for BSCS curriculum programs.

Modifications to the Learning Cycle

The following sections use the phases of the BSCS model to describe additions and modifications to the original Learning Cycle, and I make connections to the theme of creating teachable moments.

ENGAGE

This phase of the 5E Model was added to the original Learning Cycle. At a workshop on the SCIS Learning Cycle, I asked Jack Renner about the need to explicitly engage students. He suggested that the *exploration* phase of the Learning Cycle served that purpose. I still thought that a separate phase that clearly engaged students was an important means to mentally establish students' focus, expose their current understanding, possibly initiate a state of disequilibrium, and create a teachable moment. So, in our work at BSCS, we added the *engage* phase.

If you needed three words to express this phase, *creating teachable moments* would be ideal. All of the features of a teachable moment would be included—personal meaning, intellectual puzzlement, individual motivation, and readiness to learn. The exception to common descriptions of teachable moments should be obvious: Designing an engaging experience is not an unplanned experience.

EXPLORE

This phase is similar to the original *explore* phase of the Learning Cycle. In the BSCS project, we decided to use the Learning Cycle because of the strong research base, more of which is discussed in the next chapter. As an updated addition, we added an emphasis on cooperative learning based on the research of David and Roger Johnson and their colleagues (Johnson and Johnson 1987; Johnson, Johnson, and Holubec 1986).

The *explore* phase should send students in a direction that will help them begin resolving any disequilibrium of the teachable moment from the *engage* phase. It may also be the case that some students not originally engaged (i.e., they did not experience a teachable moment) will be engaged by the explore activities. The exploration allows the teacher to gain an understanding of students' knowledge of the experience and the related concepts. Students likely will have some knowledge of the concepts to be explored, but in a class of 25 students (or more), there also will likely be substantial variation in students' knowledge and understanding.

EXPLAIN

In this phase, the students try to explain what the teacher or curriculum created as they experienced in the *engage* and *explore* activities. This phase is a variation from the Learning Cycle's invention or "term introduction" phase. Note also that we used the term *explain* as a third *E* and variation from the Learning Cycle. The term *explanation* is commonly used in science literature, so it aligned well in the original use of the BSCS model. An explanation refers to the act or a process that makes an idea comprehensible.

A teachable moment expresses a time when students are motivated and open to an explanation. The teacher's challenge is to make the explanation clear, simple, succinct, direct, and understandable to the students. Vocabulary terms may be introduced and connections should be made to the *engage* and *explore* experiences.

Contemporary emphasis on *NGSS*, especially the science and engineering practices in which students develop arguments and their own explanations of concepts, highlights the need for students to try out their explanations based on what they learned from experiences in the *engage* and *explore* phases. Students may experience a new teachable moment with the challenge to use new vocabulary and ideas in ways that make them their own.

Teachers should listen to what students say and how they incorporate the new explanation. Do students' use of terms and application of concepts make sense? Students may not express complete and clear understanding of concepts and practices, but they should be able to express what they understand by using new, more accurate vocabulary and ideas.

ELABORATE

The original Learning Cycle's third phase was *discovery*. Later, in a series of 1977 publications, Karplus referred to this phase as *concept application* (Lawson, Abraham, and Renner 1989). For the BSCS model, we decided to use the term *elaborate* to capture the Piagetian perspective of what happens conceptually. Terms such as *extend* and *extension* also express the aim and what is required of the student during activities in this phase.

Some individuals have had difficulty with the term *elaborate*. I think it does describe what is intended, in a Piagetian sense—that is, an elaboration of mental structures. But the term does not clearly express what students and teachers should do, which, staying with *E* words, would most appropriately be that students *extended* their thinking.

EVALUATE

In the process of designing the instructional model, we listened to the advice of classroom teachers on our advisory board and panels to help us design specific components of the model. Teachers consistently told us that testing and assessment were not only important but required. So, we decided to include a final activity that would help teachers assess student learning. We decided to use the term *evaluate*. At the time, our idea was to incorporate another activity that would be used as an assessment. The teacher would introduce the lesson and then literally and figuratively step back from teaching and monitor the evaluation. In this way, we incorporated an embedded assessment into the 5E Model.

From a teacher's perspective, the *evaluate* phase should answer the questions, "How successful was the teaching part of creating a teachable moment?" and "How successful was student learning related to the teachable moment?" From a curriculum perspective, we later found that designing the *evaluate* activity first was an ideal application of backward design to the process of developing school programs.

Some have criticized the BSCS 5E Instructional Model because it appears that teachers do not assess student learning until the end of the sequence. That is not accurate. Teachers have opportunities embedded within each phase of the 5E Model. In the *engage* phase, they listen for current conceptions (i.e., misconceptions). In the *explore* phase, they evaluate students' process of equilibration (i.e., putting ideas together). In the *explain* phase, they assess how students' explanations have improved. And in the *elaborate* phase, they determine how well students can transfer what they have learned to a new situation.

What are the commonalities and differences between the SCIS Learning Cycle and the BSCS 5E Instructional Model? The one principal commonality underlying both models is the work of Jean Piaget (Piaget and Inhelder 1969; Piaget 1975; Bybee and Sund 1982) and subsequent research consistent with constructivist learning, specifically the focus of cognitive sciences and the work on misconceptions, the difference between novice and expert explanations of phenomena. The view of learning is summarized and discussed in greater detail in the next chapter.

The changes introduced to the Learning Cycle reflected research on learning published since the first work on the original SCIS Learning Cycle. BSCS recognized the need for an initial phase that engaged the learner in the science concepts and a concluding phase that evaluated the learner's understanding of those concepts. These phases were additions to the Learning Cycle. Beyond changes in terms describing the original Learning Cycle, embedding cooperative learning within phases was a primary modification to the Learning Cycle.

Figure 2.1 (p. 24) summarizes several historical instructional models that influenced formation of the BSCS 5E Instructional Model.

CONCLUSION

The uniqueness of the BSCS 5E Instructional Model is related to the alliterative nature of terms used to identify the model's phases. Every stage of the model begins with the same letter—*E*. When we compare this 5E Model with Herbart's (1901) model of preparation, presentation, generalization, and application or the Learning Cycle (Atkin and Karplus 1962) model of exploration, invention, and discovery, it becomes apparent why those models did not catch on among educators. A danger, of course, is that something that is catchy and easy to remember might be misused as often as it is used effectively; however, something that cannot be remembered or understood is less likely to have widespread sustainable effects.

The perspective underlying the instructional models for both SCIS and BSCS views learning as dynamic and interactive. Individuals redefine, reorganize, elaborate, and change their initial concepts and abilities through interaction with their environment and/or other individuals. The learner "interprets" objects and phenomena and internalizes the interpretation in terms of current concepts similar to the experience presented or encountered. Changing and improving conceptions often requires challenging the

Figure 2.1. Origins and Development of Instructional Models

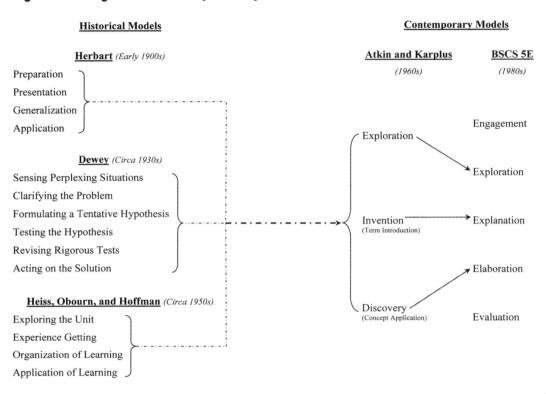

Copyright © 2006 BSCS. Image from Bybee, R. W., J. A. Taylor, A. Gardner, P. Van Scotter, J. Carlson Powell, A. Westbrook, and N. Landes. 2006. *The BSCS 5E instructional model: Origins, effectiveness and applications. www.bscs.org/bscs-5e-instructional-model*

students' current conceptions and showing those conceptions to be inadequate. If a current conception is challenged, there must be opportunities, in the form of time and experiences, to form a more adequate conception. In sum, the students' construction of knowledge can be assisted by using sequences of lessons designed to challenge current conceptions and provide time and opportunities for reconstruction to occur, and the ideal way to begin this process is with a teachable moment.

I will conclude the historical discussion with some questions: Is there a need to modify the instructional model once again? Given contemporary initiatives such as the *NGSS* and science, technology, engineering, and mathematics (STEM) education, is the BSCS 5E Instructional Model still viable? Should the model be modified in ways that accommodate current education research and social trends? At this point in the discussion, I would say no. The BSCS model can accommodate current trends and issues. Later chapters specifically discuss this response in greater detail.

REFERENCES

Abraham, M. R., and J. W. Renner. 1986. A descriptive instrument for use in investigating science laboratories. *Journal of Research in Science Teaching* 19 (2): 155–165.

Atkin, J. M., and R. Karplus. 1962. Discovery of invention? *The Science Teacher* 29 (5): 45.

Biological Sciences Curriculum Study (BSCS) and IBM. 1989. *New designs for elementary school science and health: A cooperative project between Biological Sciences Curriculum Study (BSCS) and International Business Machines (IBM).* Dubuque, IA: Kendall/Hunt Publishing Company.

Bybee, R. 1997. *Achieving scientific literacy.* Portsmouth, NH: Heinemann.

Bybee, R. W., and R. B. Sund. 1982. *Piaget for educators.* Columbus, OH: Merrill.

Bybee, R. W., J. A. Taylor, A. Gardner, P. Van Scotter, J. C. Powell, A. Westbrook, and N. Landes. 2006. *The BSCS 5E Instructional Model: Origins and effectiveness.* Colorado Springs, CO: Biological Sciences Curriculum Study (BSCS).

Commission on Secondary School Curriculum. 1937. *Science in general education.* New York: Appleton-Century-Crofts.

Compayre, G. 1907. *Herbart and education by instruction*, trans. M. Findloy. New York: Crowell.

DeBoer, G. 1991. *A history of ideas in science education.* New York: Teachers College Press, Columbia University.

Dewey, J. 1916. *Democracy and education: An introduction to the philosophy.* New York: Macmillan Company.

Dewey, J. 1933. *How we think: A restatement of the relations of reflective thinking to the educative process.* Originally published in 1910. Boston: D. C. Heath.

Dewey, J. 1938. *Experiences and education.* West Lafayette, IN: Kappa Delta Pi.

Egan, D. E., and J. G. Greeno. 1973. Piagetian theory and instruction in physics. *The Physics Teacher* 11 (3): 165–169.

Fuller, R., ed. 2002. *A love of discovery: Science education—the second career of Robert Karplus.* New York: Kluwer Academic/Plenum Publishers.

Gagne, R. M., and L. T. Brown. 1961. Some factors in the programming of conceptual learning. *Journal of Experimental Psychology* 62: 313–321.

Hawkins, D. 1965. Messing about in science. *Science and Children* 2 (5): 5–9.

Heiss, E. D., E. S. Obourn, and C. W. Hoffman. 1950. *Modern science teaching.* New York: Macmillan Company.

Herbart, J. 1901. *Outlines of educational doctrine*, trans. C. DeGarmo; ed. A. Lange. New York: Macmillan.

Howard, R. S. 1927. The unit method of instruction as applied to the teaching of physics. *School Science & Mathematics* 27 (8): 844–854.

Johnson, D. W., and R. T. Johnson. 1987. *Learning together and alone.* 2nd ed. Englewood Cliffs, NJ: Prentice Hall.

Johnson, D., R. Johnson, and E. Holubec. 1986. *Circles of learning: Cooperation in the classroom.* Alexandria, VA: Association for Supervision and Curriculum Development (ASCD).

Lawson, A. 1988. A better way to teach biology. *American Biology Teacher* 50 (5): 266–289.

Lawson, A. E. 1995. *Science teaching and the development of thinking.* Belmont, CA: Wadsworth Publishing Company.

Lawson, A. E., M. Abraham, and J. Renner. 1989. *A theory of instruction: Using the learning cycle to teach science concepts and thinking skills.* NARST Monograph, Number One. Cincinnati, OH: National Association for Research in Science Teaching (NARST).

Marek, E., and A. Cavallo. 1997. *The learning cycle: Elementary school science and beyond.* Portsmouth, NH: Heinemann.

Piaget, J. 1975. From noise to order: The psychological development of knowledge and phenocopy in biology. *Urban Review* 8 (3): 209.

Piaget, J., and B. Inhelder. 1969. *The psychology of the child.* New York: Basic Books.

Renner, J. W., M. R. Abraham, and H. H. Birnie. 1985. The importance of the form of student acquisition of data in physics learning cycles. *Journal of Research in Science Teaching* 22 (4): 303–325.

Renner, J. W., M. R. Abraham, and H. H. Birnie. 1988. The necessity of each phase of the learning cycle in teaching high school physics. *Journal of Research in Science Teaching* 25 (1): 39–58.

Renner, J. W., and A. E. Lawson. 1975. Intellectual development in pre-service elementary school teachers: An evaluation. *Journal of College Science Teaching* 5 (2): 89–92.

Roughead, W. G., and J. M. Scandura. 1968. What is learned in mathematical discovery? *Journal of Educational Psychology* 59: 283–289.

Shymansky, J. A., W. C. Kyle, and J. M. Alport. 1983. The effects of new science curricula on student performance. *Journal of Research in Science Teaching* 20: 387–404.

Stage, E. K. 2006. History of experiential learning in the U.S.: Where did the learning cycle come from? Paper presented at the GEMS Science Education Forum, Gakushuin Women's College, Tokyo, Japan.

EXPLAIN

A
Contemporary
Discussion

The BSCS 5E Instructional Model

I am convinced that the ultimate reform of education will only occur at the level of classrooms and teachers, so I am recommending an instructional model that touches the heart of teaching—the most practical level at which education reform occurs.

This chapter presents a detailed explanation of the BSCS 5E Instructional Model. The BSCS 5E Instructional Model, commonly referred to as the 5Es, consists of *engage*, *explore*, *explain*, *elaborate*, and *evaluate*. Each phase has a specific purpose and contributes to the teacher's coherent instruction and the students' constructing a better understanding of content, attitudes, and skills. The model can be used to help frame an integrated instructional sequence that may be a unit or an entire program. Once internalized, the model also can inform many of the instantaneous decisions teachers must make in the flow and dynamics of classrooms.

PSYCHOLOGICAL FOUNDATIONS

The BSCS 5E Instructional Model is grounded in the psychology of learning, specifically in a constructivist perspective. Understanding how students learn seems to be an appropriate introduction. From there, this chapter progresses to a detailed explanation of the phases and implications for educational materials and classroom instruction.

How Students Construct Concepts

Intimate is not too strong a word to describe the relationship between how students learn and the way teachers should teach. There should be congruence between teaching strategies and a model of learning. In the case of the 5E Model, constructivism is the perspective most important for teachers to understand.

Constructivism is the psychological foundation for the BSCS 5E Instructional Model. The constructivist view assumes a dynamic and interactive conception of human learning. Students bring their current explanations, attitudes, and skills to a learning experience. Through meaningful interactions with other individuals and their physical environment, which includes students and teachers, they redefine, replace, reorganize, and reconstruct initial explanations, attitudes, and skills. In developing the instructional model, I assumed that the constructive process could be assisted by sequences of experiences designed to challenge students' current conceptions, attitudes, and skills, and that provided time and opportunities for reconstruction to occur.

You might ask, How does this process of learning actually occur? I will briefly digress to answer the question by describing a process of equilibration. This model of learning was described by the psychologist Jean Piaget. Piaget's model of equilibration is the part of his overall developmental theory that relates to the process of learning. Equilibration is, in essence, the model of learning in Piaget's developmental psychology. Piaget's major statement on equilibration is *The Development of Thought* (1975).

According to Piaget's model, intellectual development occurs through an adaptation of cognitive structures in response to a discrepancy between the individual's current cognitive structure and a cognitive referent (e.g., an object or event) in the environment. Disequilibrium results from such a discrepancy. Modification of intellectual structures brings the cognitive system back to equilibrium. The process of equilibration involves both maintenance and change of intellectual structures—that is, old structures continue while they are modified or combined with others to form new structures.

Here is a simple example of the simultaneous maintenance and change of cognitive structures. Imagine an infant's first encounter with a ball—in this case, let us say a Ping-Pong ball. The infant's interaction with this Ping-Pong ball formulates a cognitive structure of *ball*. Later, the infant encounters a golf ball. It is white and about the same size as the Ping-Pong ball and has many similar properties, such as that it is white and bounces if dropped. But it is not smooth, and it is heavier than the Ping-Pong ball. This second experience likely causes some disequilibrium and the cognitive reconstruction of this infant's conception of a ball. You can continue this example by considering a tennis ball and, ultimately, balls that are quite dissimilar to the original experience, such as a basketball or football. Across time, the interactions with the different balls results in an elaborate concept of *ball* and characteristics that defined "ballness." Equilibration was the process that resulted in the maintenance and change from the original cognitive structure to a more complex and elaborate concept of balls.

Equilibration occurs at three major levels: (1) the total system (which Piaget described as a stage of development), a complete set of cognitive structures that are at equilibrium for a period of time; (2) a subsystem, such as the cognitive structures for classification, conservation, and numeration; and (3) specific concepts, such as color, objects, and organisms, which may be included in the aforementioned levels of organization.

In Piaget's model, organization and adaptation are the two processes that bring about equilibration. Organization is the maintenance of an internal order of the intellectual structure through the inherent tendency to systematize and integrate intellectual structures into coherent systems. This tendency results in the spontaneous transition to higher orders of intellectual complexity while maintaining original concepts. Think of the ball example.

Adaptation is the process of changing the intellectual structure through interaction with the environment. This modification results in development of the cognitive structure.

Intellectual adaptation consists of two processes that are simultaneous and complementary. These processes are *accommodation* and *assimilation*.

I can clarify the processes of accommodation and assimilation with another example. This example also may be seen as a teachable moment(s) and the ensuing psychological processes. Assume that a student is at equilibrium cognitively with respect to particular events in the world—say, the floating of ice in water; that is, based on past experiences, the student has a conception that satisfactorily explains why ice floats in water. The teacher confronts the student with a puzzling demonstration—a discrepant event. For example, the teacher releases an ice cube in a clear liquid and it sinks. This instance of an ice cube sinking apparently contradicts prior experiences of ice cubes floating. The student tries to explain the event in terms of prior experiences (assimilation) and cannot do so. So the student has to change some cognitive schemes to explain the anomalous event. Accommodation occurs in the same time frame as the assimilation. The intellectual structure is somewhat readjusted to the external reality, the phenomenon of the demonstration. The student's prior conception of how ice floats or sinks is changed, or the student develops a new expla-

nation that maintains the prior "ice floats in water" conception. For instance, the student explains the event using another model (the liquid is not water, so the ice would not float—it would sink). The prediction graph in Figure 3.1 shows how a student can develop an understanding of a concept based on what they observe.

I briefly summarized the process of equilibration for several reasons. Much of the recent research on the cognitive sciences incorporates a similar model. The BSCS 5E Instructional Model assumes that students construct knowledge, attitudes, and skills in a similar manner.

The model of learning proposed by Piaget influenced a long line of research, with implications for science teaching in the 1980s (see, for example, Hewson and Hewson 1988). In the 1990s, based on the work of Lev Vygotsky (1978), early ideas about constructivism were adapted to include social interaction and the role of language; Vygotsky's work

Figure 3.1. Prediction Graph Showing Student Understanding

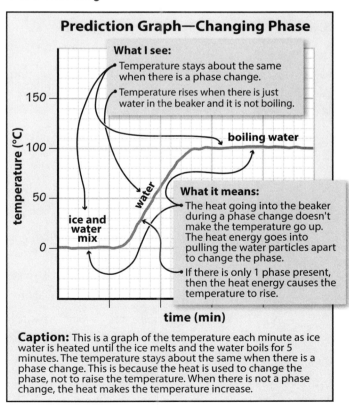

Caption: This is a graph of the temperature each minute as ice water is heated until the ice melts and the water boils for 5 minutes. The temperature stays about the same when there is a phase change. This is because the heat is used to change the phase, not to raise the temperature. When there is not a phase change, the heat makes the temperature increase.

contributed to an understanding of the social context in cognitive development (Hodson and Hodson 1998).

Vygotsky also made another significant contribution as part of criticism of strict interpretations of the Piagetian view. Vygotsky proposed that individuals have a "zone of proximal development," which he described as the distance between the actual developmental level as determined by independent problem solving and the level of potential development as determined through problem solving under adult guidance or in collaboration with capable peers (Vygotsky 1978, p. 86).

Based on the importance of social interaction and the students' zone of proximal development, one sees implications for the role of group work in activities and investigations, the design of instructional materials, and the teacher's guidance in the process of learning.

This line of research on learning has support in National Research Council reports such as *How People Learn: Brain, Mind, Experience, and School* (Bransford, Brown, and Cocking 1999); *How Students Learn: Science in the Classroom* (Donovan and Bransford 2005); and *How People Learn: Bridging Research and Practice* (Donovan, Bransford, and Pellegrino 1999).

DESIGNING AN INSTRUCTIONAL MODEL

As teachers think about learning and instruction, it is common to focus on single activities or lessons. I recommend expanding this perception to an integrated instructional sequence (NRC 2006) and an instructional model such as the 5Es. The challenge for educators requires translating education research to instructional practice. The model has to have evidence-based strategies and a sequence that is described in practical terms for teachers. Having features that are understandable by students would be an added benefit.

Design Specifications

The following statements describe some of the specifications for an instructional model. In broad terms, the model should be grounded in a synthesis of contemporary research on the psychology of learning—*How People Learn* (Bransford, Brown, and Cocking 1999).

- The model must have an initial stage that engages the learners and brings about disequilibrium in the students. This stage should be designed to gain students' attention on critical aspects of an experience, focus students' thinking on concepts deemed important, and expose any misconceptions.

- The model must help learners integrate new knowledge and abilities with current knowledge and abilities. To understand any new concept or apply a new ability, learners must connect a concept or ability to other concepts and abilities in a meaningful way.

- The model must allow for student-student as well as student-teacher interaction. Social interaction among students encourages argument and opportunities to formulate new explanations for common experiences.

- The model must provide time and opportunities in the form of experiences to accommodate the process of conceptual change. That is, the educational activities must be applicable to a range of students' knowledge and abilities. Designing an instructional model unique to each concept seems complex and inefficient and fails to recognize the fact that teachers must interact with 25 or more students at a time. This specification is both crucial and challenging. The research on conceptual change necessarily focuses on specific concepts, but teachers cannot anticipate and recognize all students' conceptions associated with a particular topic or curriculum. The model must therefore allow for explanations generated by students as well as ample time for discussion of those explanations. The model must allow for explanations generated by students and not just the teacher giving students a clear, coherent explanation.

- While allowing time and opportunities for conceptual change, the model must be manageable by the teacher. It is generally not feasible for a teacher to ascertain each student's prior conceptions and then structure instruction accordingly. In the future, technology may help overcome this problem. But for now, teachers will confront classrooms where students present considerable variation in their conceptual understanding. Thus, the model should include strategies that are designed to maximize the benefit for individual students and also that are usable for the teacher in the typical classroom environment.

- The model must be understandable. Teachers who initially use an instructional model likely are themselves engaged in a process of conceptual change. Most teachers already have models of teaching and learning that probably vary with the one presented here. The new model should therefore be described in terms that are understandable. Students also should be able to identify what phase of the model they are in and what is expected of them during that phase so they can engage appropriately.

- The model should accommodate a variety of teaching strategies and activities that may include web-based searches, computer simulations, reading, laboratory investigations, writing, direct instruction, and guided inquiry.

- The model must provide opportunities for informal as well as formal assessments of student learning.

In summary, the instructional model should initially engage the learner, challenge current conceptions and abilities, recognize the value of social interaction, provide time and opportunities for conceptual change, be manageable and understandable, accommodate a variety of instructional strategies, and have embedded opportunities for informal and formal assessments. The BSCS 5E Instructional Model meets these criteria.

THE BSCS 5E INSTRUCTIONAL MODEL: A CONTEMPORARY DISCUSSION

Beginning in the late 1980s, the Biological Sciences Curriculum Study (BSCS) implemented an instructional model in most of its programs. That model is referred to as the BSCS 5E Instructional Model and consists of the following phases: *engage, explore, explain, elaborate,* and *evaluate*. Each phase has a specific function and contributes to the teacher's coherent instruction and the students' formulating a better understanding of the knowledge, attitudes, and skill emphasized in the instructional sequence. The model has been used to help frame the sequence and organization of programs, units, and lessons. Once internalized by the teacher, it also can inform the many instantaneous decisions one must make in classroom situations.

Curriculum developers and classroom teachers can apply the model at several levels. The model can be the organizational pattern for a yearlong program, for units within the curriculum, and for sequences of lessons occurring over a period longer than one class period. I will note that the model's intent is conceptual development. Thus, the phases will likely be implemented over more than a class period. My primary recommendation is to use the BSCS 5E Instructional Model at the following levels: yearlong programs, units, and chapters, not as daily lessons.

I recommend, however, that developers and teachers initially use the model clearly and consistently at only one of these levels, preferably for a unit-level instructional sequence. Although curriculum developers who are familiar with the model may be able to use nested cycles (cycles within cycles) or interwoven cycles (exploring one concept while elaborating on another), apprising teachers and students of that nesting or interweaving would neglect the "intelligible and apparent" criterion.

The BSCS 5E Instructional Model is built on a constructivist view of learning; you may recall that constructivism represents an interactive model of how humans learn. Using this approach, students redefine, reorganize, elaborate, and change their initial concepts and abilities through self-reflection and interaction with their peers, teachers, and environments. Learners interpret objects and phenomena and internalize those interpretations in terms of their current conceptional understanding. The objective in a constructivist program is to challenge students' current conceptions by providing discrepant events (i.e., teachable moments), data that conflict with students' current thinking, or experiences that provide an alternative way of thinking about objects and phenomena. When an activity challenges students' conceptions, there must be an opportunity for students to reconstruct a

Teachers learning the BSCS 5E Model.
Copyright © 2014 BSCS. Image from Diabetes Education Curriculum Grades K–12 Professional Development Workshop.

conception that is more adequate than their original conception. This takes time, a planned sequence of instruction, and a teacher's judgment about students' capacity to adapt to the new concepts and abilities. The instructional model outlined in the next sections provides the time, opportunity, and structure necessary for such learning to occur.

The following sections present descriptions of the five phases of the instructional model. Curriculum developers and classroom teachers can apply the instructional model in different disciplines and with varied teaching strategies.

Engage: Creating a Teachable Moment

The purpose of this phase is to generate interest, stimulate curiosity, and raise questions. The first phase should engage the student in the learning task. From a teacher's perspective, you are trying to create a teachable moment for the students. The student mentally focuses on an object, problem, situation, or event. The activities of this phase should help students make connections to past and future activities. The connections depend on the learning task and may be conceptual, procedural, or behavioral.

Asking a question, defining a problem, showing a discrepant event, and acting out a problematic situation are all ways to engage students and allow them to focus on the instructional activities. The teacher's role is to present a situation and identify the instructional task. The teacher also sets the rules and procedures for the activity. The experience need not be long or complex; in fact, it should probably be short and simple.

Successful engagement results in students being puzzled by, and actively motivated by, the learning activity. Here, the word *activity* refers to both a constructivist and a behavioral approach—that is, the students are mentally and physically active.

If students look perplexed or bewildered, expressing, "What happened?" or "Wow, I was surprised by that," and "I can't explain that," they likely are engaged in a learning situation. You have created a teachable moment. Students have some ideas, but the expression of concepts and use of their abilities may not be scientifically accurate and productive.

Over the decades, I have come to realize two things about this phase. The engagement need not be a full lesson, but usually it is a full lesson because of the need to surface and assess students' current knowledge. It might be as brief as posing a question or showing a short demonstration, followed by discussion. Teachers might, for example, provide a brief description of natural phenomenon and ask how students would explain the situation. The main point is that the students are puzzled and thinking about content related to the learning outcomes of the instructional sequence. The second point about this phase and the next is that they present opportunities for teachers to informally determine misconceptions expressed by the students. I emphasize the informal nature of these observations. The *engage* phase is not a formal preassessment.

Explore: Recognizing Students' Current Understandings and Abilities

Once you have engaged students, they need time to experience related phenomenon and examine their ideas. Exploration activities are designed so that all students have common, concrete experiences on which they continue building concepts, processes, and skills. If engagement brings about disequilibrium, exploration initiates the process of equilibration. This phase should be concrete and meaningful as the students investigate questions and problems and propose answers and solutions.

The aim of exploration activities is to establish experiences that teachers and students can use later to formally introduce and discuss scientific and technological concepts, practices, or skills. During the activity, the students have time in which they can explore objects, events, or situations. As a result of their mental and physical involvement in the activity, the students establish relationships, observe patterns, identify variables, and question events.

The teacher's role in the *explore* phase is that of facilitator or coach. The teacher initiates the activity and allows the students time and opportunity to investigate objects, materials, and situations based on each student's own ideas of the phenomena. If called upon, the teacher may coach or guide students as they begin constructing new explanations. Use of tangible materials and concrete experiences is essential in the exploration phase.

A portion of the exploration phase should center on group work and social interaction or cooperative learning. The opportunity for students to interact, discuss, and argue in a constructive environment and focus on goal-centered activities enhances the possibility that their current concepts will be challenged and other ideas will be evident as they reconstruct their ideas.

Explain: Developing New Understandings and Abilities

The term *explanation* means the act or process in which concepts, processes, or skills become plain, comprehensible, and clear. The purpose of this phase includes introducing concepts and practices that may be used to interpret experiences and propose explanations. The process of explanation provides the students and teachers with a common use of terms relative to the learning experience. In this phase, the teacher directs student attention to specific aspects of the engagement and exploration experiences. First, the teacher asks the students to give their explanations. Second, the teacher introduces scientific or technological explanations in a succinct, direct, and formal manner. Explanations are ways of ordering and giving a common language for the exploratory experiences. The teacher should base the initial part of this phase on the students' explanations and clearly connect the explanations to experiences in the *engage* and *explore* phases of the instructional model.

Teachers applying the BSCS 5E Model.
Copyright © 2014 BSCS. Image from BSCS ViSTA Plus Professional Development Workshop.

The key to this phase is to present concepts, processes, or skills briefly, simply, clearly, and directly, then continue on to the next phase.

In the *explain* phase, teachers have a variety of techniques and strategies at their disposal. Educators commonly use verbal explanation, but there are numerous other strategies, such as video, reading, the web, and educational software. This phase continues the process of mental ordering and provides vocabulary for concepts and abilities. In the end, students should be able to explain their experiences to each other and to the teacher.

I would make the point that the role of the teacher in this phase shifts from helping students explore questions and ideas to helping them make meaning of their experiences. This also is an opportunity for the teacher to provide students with information to help them connect their prior experiences with explaining phenomena. Verbal explanations or formal teaching are common, but use of other means may enhance students' understanding and abilities.

Elaborate: Transferring Concepts and Abilities to New Situations

The purpose of this phase centers on the use and application of concepts and explanations in new contexts. Once the students have an explanation of their learning tasks, it is important to involve them in further experiences that apply, extend, or elaborate the concepts, processes, or skills. Some students may still have misconceptions, or they may only understand a concept in terms of the exploratory experience. Elaboration activities provide further time and experiences that contribute to learning. The intention is to facilitate the transfer of concepts and abilities to related but new situations. A key point for this phase: Use activities that are a new challenge but achievable by the students, and have students use different modes for explanations, such as written and oral presentations, diagrams, graphs, and mathematics.

In the *elaborate* phase, the teacher encourages interactions among students in groups and with other sources, such as written material, databases, simulations, and web-based searches. The teacher should clarify the groups' task and set the stage for discussions where students present and defend their approaches and proposed results to the new situation. As a result of their work, students expand and deepen their understandings and abilities. The use of interactions within student groups as a part of the elaboration process is

Teachers considering a phase of the BSCS 5E Model.

important. Group discussions and cooperative learning situations provide opportunities for students to express their understanding of the subject and receive feedback from others who are very close to their own level of understanding.

The *elaborate* phase also is an opportunity to involve students in new situations and problems that require the application of identical or similar explanations. Transfer of concepts, processes, and skills is an important goal of the *elaborate* phase.

Evaluate: Assessing Students' Understandings and Abilities

Teachers recognize the importance of students' receiving feedback on the adequacy of their explanations and abilities. To be clear, informal evaluation occurs from the beginning of the teaching sequence. The teacher can complete a formal evaluation after the *elaborate* phase. As a practical educational matter, teachers must evaluate education outcomes. This is the phase in which teachers administer assessments to determine each student's level of understanding.

In the *evaluate* phase, the teacher should involve students in experiences that are understandable and consistent with those of prior phases and congruent with the explanations. The teacher should first determine the evidence for student learning, then identify the means of obtaining that evidence, as part of the *evaluate* phase. Tables 3.1 (pp. 40–41) and 3.2 (pp. 41–42) summarize the BSCS 5E Instructional Model with examples of teacher and student behaviors.

Contemporary Reflections on the BSCS 5E Instructional Model

In the 1980s and 1990s, as part of my work at BSCS, I used this instructional model in the design of curriculum materials for elementary, middle, and high school. In subsequent years, I have applied the 5E Model to a variety of other instructional programs and contemporary policies. Here are some reflections on the 5E Model and my experiences.

My first reflection centers on the theme of creating teachable moments and the personal engagement and interest of objects and events for students. Obviously, one should ask, What is the potential personal meaning of an activity for a particular phase of instruction?

Students derive personal meaning from three types of experiences. The general terms *physical, psychological*, and *social* describe the types of meaning an experience has for students. An activity can have meaning for students because objects and events are physically close. Objects and events have meaning for individuals simply because they are close and involved. Placing an unknown object in a student's hand increases the meaning of that object for the student. Having hands-on experiences and engaging in problem-solving activities both have a dimension of personal meaning.

There is a second dimension of psychological meaning. Some objects and events are interesting and engaging for students. Dinosaurs, plants, weather, and the solar system

Table 3.1. The BSCS 5E Instructional Model: What the Teacher Does

STAGE OF THE INSTRUCTIONAL MODEL	WHAT THE TEACHER DOES THAT IS ...	
	CONSISTENT WITH THIS MODEL	**INCONSISTENT WITH THIS MODEL**
Engage	• Creates interest • Generates curiosity • Raises questions • Elicits responses that uncover what the students know or think about the concept or topic	• Explains concepts • Provides definitions and answers • States conclusions • Provides closure • Involves lectures
Explore	• Encourages the students to work together without direct instruction from the teacher • Observes and listens to the students as they interact • Asks probing questions to redirect the students' investigations when necessary • Provides time for the students to puzzle through problems • Acts as a consultant for students	• Provides answers • Tells or explains how to work through the problem • Provides closure • Tells the students that they are wrong • Gives information or facts that solve the problem • Leads the students step by step to a solution
Explain	• Encourages the students to explain concepts and definitions in their own words • Asks for justification (evidence) and clarification from students • Formally provides definitions, explanations, and new labels • Uses students' previous experiences as the basis for explaining concepts	• Accepts explanations that have no justification • Neglects to solicit the students' explanations • Introduces unrelated concepts or skills
Elaborate	• Expects the students to use formal labels, definitions, and explanations provided previously • Encourages the students to apply or extend the concepts and skills in new situations • Reminds the students of alternate explanations • Refers the students to existing data and evidence and asks, "What do you already know?" "Why do you think *x*?" (Strategies from Explore also apply here.)	• Provides definitive answers • Tells the students that they are wrong • Involves lectures • Leads students step by step to a solution • Explains how to work through the problem

Table 3.1. (*continued*)

STAGE OF THE INSTRUCTIONAL MODEL	WHAT THE TEACHER DOES THAT IS …	
	CONSISTENT WITH THIS MODEL	INCONSISTENT WITH THIS MODEL
Evaluate	• Observes the students as they apply new concepts and skills • Assesses students' knowledge and/or skills • Looks for evidence that the students have changed their thinking or behaviors • Allows students to assess their own learning and group-process skills • Asks open-ended questions such as, "Why do you think … ?" "What evidence do you have?" "What do you know about x?" "How would you explain x?"	• Tests vocabulary words, terms, and isolated facts • Introduces new ideas or concepts • Creates ambiguity • Promotes open-ended discussion unrelated to the concept or skill

Table 3.2. The BSCS 5E Instructional Model: What the Student Does

STAGE OF THE INSTRUCTIONAL MODEL	WHAT THE STUDENT DOES THAT IS …	
	CONSISTENT WITH THIS MODEL	INCONSISTENT WITH THIS MODEL
Engage	• Asks questions such as, "Why did this happen?" "What do I already know about this?" "What can I find out about this?" • Shows interest in the topic	• Asks for the "right" answer • Offers the "right" answer • Insists on answers or explanations • Seeks one solution
Explore	• Thinks freely, within the limits of the activity • Tests predictions and hypotheses • Forms new predictions and hypotheses • Tries alternatives and discusses them with others • Records observations and ideas • Suspends judgment	• Lets others do the thinking and exploring (passive involvement) • Works quietly with little or no interactions with others (only appropriate when exploring ideas or feelings) • "Plays around" indiscriminately with no goal in mind • Stops with one solution
Explain	• Explains possible solutions or answers to others • Listens critically to others' explanations • Questions others' explanations • Listens to and tries to comprehend explanations that the teacher offers • Refers to previous activities • Uses recorded observations in explanations	• Proposes explanations from "thin air" with no relationship to previous experiences • Brings up irrelevant experiences and examples • Accepts explanations without justification • Does not attend to other plausible explanations

Table 3.2. (*continued*)

STAGE OF THE INSTRUCTIONAL MODEL	WHAT THE STUDENT DOES THAT IS ...	
	CONSISTENT WITH THIS MODEL	**INCONSISTENT WITH THIS MODEL**
Elaborate	• Applies new labels, definitions, explanations, and skills in new but similar situations • Uses previous information to ask questions, propose solutions, make decisions, and design experiments • Draws reasonable conclusions from evidence • Records observations and explanations • Checks for understanding among peers	• "Plays around" with no goal in mind • Ignores previous information or evidence • Draws conclusions from "thin air" • Uses only those labels that the teacher provided in discussions
Evaluate	• Answers open-ended questions by using observations, evidence, and previously accepted explanations • Demonstrates an understanding or knowledge of the concept or skill • Evaluates his or her own progress and knowledge • Asks related questions that would encourage future investigations	• Draws conclusions without using evidence or previously accepted explanations • Offers only yes or no and memorized definitions or explanations as answers • Fails to express satisfactory explanations in his or her own words • Introduces new, irrelevant topics

Source: BSCS 2013. All rights reserved. Reproduced by permission.

are all examples of psychologically interesting things for children. Instruction can use the initial interest in these areas to develop concepts such as time, cycles, and scale.

Finally, there is a social aspect of meaning. This is the dimension that most individuals associate with meaning. Educators often equate meaning with relevance or timeliness of issues. In some cases, science-related social issues—such as environmental quality, resource use, or climate change—are meaningful to students. However, assuming these are meaningful to students just because they are timely, and even critically important, is not always a sound assumption from a learning point of view. Obviously, combining all three dimensions of meaning in educational experiences certainly enhances the possibilities of students' learning.

My second issue regards the possibilities and limits of an instructional model. An instructional model is probably necessary but not sufficient for the process of conceptual change. That is, the careful structuring and sequencing of activities helps tremendously to bring about conceptual change. There remains, however, a critical interaction among students and between teachers and students that completes the process in the context of classrooms and school. A carefully structured sequence of activities enhances the possibilities of learning, but it does not ensure learning. Teachers' careful probing, subtly challenging the students, and knowing when to provide a hint or clue that will help the student reconstruct an idea are all interpersonal dimensions of instruction that cannot be adequately accommodated by a set of activities. In short, the burden and complexity of learning is too heavy to place solely on an instructional model. The teacher is essential to complete the process of conceptual change.

The instructional model provides a psychological and educational bridge from the students' current conceptions or misconceptions to appropriate scientific concepts. You may note the careful structuring and sequencing that allows students to identify explanations and their adequacy. Although the process honors students' ideas, curriculum developers and classroom teachers also need to introduce and help students learn adequate scientific concepts and processes. The design of the 5Es is such that this introduction occurs at the midpoint of the instructional sequence. After the *elaborate* and *evaluate* phases, provide time and opportunity for students and teachers to assess their own understanding against those of accepted science concepts and abilities.

My third point concerns assessment. The No Child Left Behind (NCLB) legislation has greatly influenced the form and function of assessment in American education. The fact that the 5E Model includes an *evaluate* phase is a positive and constructive response to one recommendation from a report from the Gordon Commission on the Future of Assessment in Education. To paraphrase, the report recommended that assessment resources and tools should better integrate with teaching and learning in classrooms. This is the essence of the *evaluate* phase. The assessment looks very much like good instruction in the prior phases of the model. The evaluation tasks present students with situations that engage the knowledge and practices developed in prior phases of the 5E Model.

I mention the *evaluate* phase because of teachers' concerns about assessment and accountability. Using the 5E Model and recognizing the *evaluate* phase as essential will help teachers address contemporary concerns about assessment.

Finally, I will note four general factors that support the BSCS 5E Instructional Model: (1) education research on learning in general and conceptual change in particular, (2) congruence of the model with the general processes of scientific inquiry and technological design, (3) utility of the model for designing and developing curriculum materials, and (4) practical use by classroom teachers.

CONCLUSION

This chapter describes the BSCS 5E Instructional Model originally designed for BSCS programs. An instructional model brings coherence to different teaching strategies, provides connections among educational activities, helps teachers make decisions about interactions with students, and contributes to students' development of knowledge and abilities.

The instructional model has constructivism as a theoretical foundation but recognizes the critical role of classroom teachers, who must make myriad decisions about their students.

The instructional model consists of five phases designed to facilitate the process of conceptual change: *engage, explore, explain, elaborate,* and *evaluate.* The actual application of the phases in curriculum and teaching may not be as clear and easy as outlined here; still, the model should contribute to better, more consistent, and more coherent instruction.

REFERENCES

Bransford, J., A. Brown, and R. Cocking, eds. 1999. *How people learn: Brain, mind, experience, and school.* Washington, DC: National Academies Press.

Donovan, M., and J. Bransford, eds. 2005. *How students learn: Science in the classroom.* Washington, DC: National Academies Press.

Donovan, M., J. Bransford, and J. Pellegrino. 1999. *How people learn: Bridging research and practice.* Washington, DC: National Academies Press.

Hewson, P., and M. Hewson. 1988. An appropriate conception of teaching science: A view from studies of science learning. *Science Education* 72 (5): 597–614.

Hodson, D., and J. Hodson. 1998. From constructivism to social constructivism: A Vygotskian perspective on teaching and learning science. *School Science Review* 79 (289): 33–41.

National Research Council (NRC). 2006. *America's lab report: Investigations in high school science.* Washington, DC: National Academies Press.

Piaget, J. 1975. *The development of thought.* New York: Viking Press.

Vygotsky, L. S. 1978. *Mind in society: The development of higher psychological processes.* Cambridge, MA: MIT Press.

ELABORATE

Expanding and Adapting Your Understanding

Reviewing Education Research Supporting Instructional Models

Does education research support the ideas underlying the theme of this book—creating teachable moments and the BSCS 5E Instructional Model? The general answer to this question is yes. That said, some approaches had their origin in philosophies of education. John Dewey's ideas were closely associated with science processes, but he did not conduct research supporting his applications to instruction. As you will see, the Science Curriculum Improvement Study (SCIS) Learning Cycle had a number of studies supporting that model.

Relative to the BSCS model, we did not have time or support to conduct research specifically on the 5E Model. However, the field testing of BSCS programs provided support for the instructional model. At BSCS, we took an engineering perspective. We began with a problem—we needed an instructional model—and the evidence-based Learning Cycle provided what I deemed a reasonable and prudent initial solution to the problem. The BSCS team made changes and additions to the Learning Cycle. However, in the decades since the BSCS team developed and implemented the instructional model, research has supported the model's efficacy.

Historically, educators have explained learning in one of three broad categories: transmission, maturation, and construction. B.F. Skinner's theories of reinforcement serve as an example of the first category—transmission (Ferster and Skinner 1957). Arnold Gesell's model exemplifies the second—maturation (Ilg and Ames 1955). Many educators would recognize Piaget's ideas as associated with the third category—construction (Piaget 1975; Piaget and Inhelder 1969). Piaget viewed development of the intellect as neither direct learning from the environment (i.e., reinforcement) nor maturation. Rather, he proposed that learning consists of reorganization and reconstruction of cognitive structures as a result of interactions between the individual and the environment. Although simplified, this description provides an orientation for discussions in this chapter. The three views of learning are summarized in Table 4.1 (p. 48).

Table 4.1. Views of Learning, Knowledge, and Teaching

PERSPECTIVE	ASSUMPTIONS ABOUT STUDENTS AS LEARNERS	VIEWS OF KNOWLEDGE	APPROACHES TO INSTRUCTION
Transmission	Students can be "filled" with facts, information, and concepts.	Concepts are a copy of reality.	External to internal: Present knowledge directly.
Maturation	Students must be allowed to mature and develop.	Concepts emerge with age.	Internal to external: Allow knowledge and understanding to develop.
Construction	Students should be actively involved in learning.	Concepts are constructed from experiences.	Interactions between internal cognitive structures and external experience result in knowledge and the acquisition and construction of concepts and abilities.

These are simplified views. Experienced teachers realize that all three have some support in research and no single view explains learning in classrooms. The brief discussion and summary table does, however, locate the orientation for the BSCS 5E Instructional Model. It is best described as based on the constructive/interactive view of learning.

RESEARCH ON LEARNING AND INSTRUCTION

The BSCS 5E Instructional Model builds on the work of other instructional models and is supported by current research on learning. BSCS has a long history of developing curriculum materials that reflect the most recent research about learning and teaching. Our current understanding has been informed by research conducted by cognitive scientists from around the world (Brooks and Brooks 1993; Driver, Asoko, et al. 1994; Driver, Squires, et. al 1994; Lambert et al. 1995; Matthews 1992; NRC 2000; Piaget 1976; Posner et al. 1982; Vygotsky 1962). Cognitive research shows that learning is an active process occurring within and influenced by the learner. Hence, learning results from an interaction between experiences and how the student processes those experiences based on perceived notions and extant personal knowledge. The BSCS 5E Instructional Model applies this research to curriculum materials.

The 5E Model had its origins with the work of others, especially the SCIS Learning Cycle. Research indicated the effectiveness of the original Learning Cycle:

- All three phases of the model must be included in instruction, and the *explore* phase must precede the invention (term introduction) phase.

- The specific instructional format may be less important than including all phases of the model, but laboratory work (typical in the *explore* phase) is more effective for many students, provided it is followed by invention (term introduction).

- Finally, student attitudes toward science instruction are more positive when they are allowed to explore concepts through experimentation or other activities before discussing them (Lawson, Abraham, and Renner 1989).

How Students Learn

Several reports from the National Research Council (NRC) and the National Academies of Science (NAS) have synthesized contemporary research on learning. The first NRC report, *How People Learn: Brain, Mind, Experience, and School* (Bransford, Brown, and Cocking 1999), was followed by other reports that go beyond the synthesis and discuss strategies for applying the findings to practice, including *How Students Learn: History, Mathematics, and Science in the Classroom* (Donovan and Bransford 2005).

How People Learn (Bransford, Brown, and Cocking 1999) offers insights about learners and learning that are especially important for this book. Three major findings summarized in *How People Learn: Bridging Research and Practice* (Donovan, Bransford, and Pellegrino 1999) are highlighted because they have a strong research base and clear implications for the use of systematic and carefully designed instruction:

1. Students come to the classroom with preconceptions about how the world works. If their initial understanding is not engaged, they may fail to grasp new concepts and information that are taught, or they may learn for the purpose of a test but revert to their preconceptions outside the classroom.

2. To develop competence in an area of inquiry, students must (a) have a deep foundation of factual knowledge, (b) understand facts and ideas in the context of a conceptual framework, and (c) organize knowledge in ways that facilitate retrieval and application.

3. A "metacognitive" approach to instruction can help students learn to take control of their own learning by defining learning goals and monitoring their progress in achieving them. (Donovan, Bransford, and Pellegrino 1999)

I direct the reader's attention to the following points that apply to the themes of this book. The first finding highlights the importance of engaging students in the learning task—that is, create a teachable moment—thus providing an opportunity to identify any preconceptions about how the world works.

The teachable moment described in the *engage* phase should also help teachers assess students' preconceptions related to the experiences and contexts related to the teachable moment.

The second point underscores the complementary and organizational aspects of factual and conceptual knowledge. I note this to counter extreme positions that would not include or emphasize facts at the expense of major conceptual ideas of a discipline. An effective education requires both facts and concepts. In addition, they need to be organized to be retrieved and applied.

Finally, there is a need for students to learn how to learn. In the 5E Model, the *elaborate* phase provides an opportunity for students to become more metacognitive as they learn to take control of their own learning.

These findings have parallel implications for classroom instruction and curriculum development. The findings imply that teachers must be able to do the following:

- Recognize and draw out preconceptions from their students and base instructional decisions on the information they get from their students
- Teach their subject matter in depth so facts are conveyed in a context with examples and a conceptual framework
- Integrate metacognitive skills into the curriculum and teach those skills explicitly

Relative to this review and the BSCS 5E Instructional Model, a quote from *How People Learn* (Bransford, Brown, and Cocking 1999) seems especially germane:

> An alternative to simply progressing through a series of exercises that derive from a scope and sequence chart is to expose students to the major patterns of a subject domain as they arise naturally in problem situations. Activities can be structured so that students are able to explore, explain, extend, and evaluate their progress. Ideas are best introduced when students see a need or a reason for their use—this helps them see relevant uses of the knowledge to make sense of what they are learning. (p. 127)

This quote directs attention to a research-based recommendation for a structure and sequence of instruction that exposes students to problem situations (i.e., engage their thinking by creating teachable moments) and then provides opportunities to explore, explain, extend, and evaluate their learning. This research summary from the NRC supports the design and sequence of the BSCS 5E Instructional Model. It even uses terms very similar to the 5Es!

Integrated Instructional Units

Following the work of Bransford, Brown, and Cocking, the NRC published *America's Lab Report: Investigations in High School Science* (2006). This report examined the status of science laboratories and developed a vision for their future role in high school science education. The NRC committee used the following definition for laboratory experiences (NRC 2006):

> *Laboratory experiences provide opportunities for students to interact directly with the material world (or with data drawn from the material world), using tools, data collection techniques, models, and theories of science. (p. 31)*

Note that this definition includes physical manipulation of materials, interactions with simulations, interactions with actual (not artificially created) data, analysis of large databases, and remote access to instruments and observations, such as via web links.

In the analysis of laboratory experiences, the NRC committee applied results from the large and still-growing body of cognitive research. Researchers had investigated the sequence of science instruction, including the role of laboratory experiences, as these sequences enhance student achievement of the learning goals. The NRC committee (NRC 2006) proposed the phrase *integrated instructional units*:

> *Integrated instructional units interweave laboratory experiences with other types of science learning activities, including lectures, reading, and discussion. Students are engaged in forming research questions, designing and executing experiments, gathering and analyzing data, and constructing arguments and conclusions as they carry out investigations. Diagnostic, formative assessments are embedded into the instructional sequence and can be used to gauge the students' developing understanding and to promote their self-reflection on their thinking. (NRC 2006, p. 82)*

Integrated instructional units have two key features. First, laboratory and other experiences are carefully designed or selected on the basis of what students should learn from them. Second, the experience is explicitly linked to and integrated with other learning activities in the unit.

The features of integrated instructional units map directly to the BSCS 5E Instructional Model. Stated another way, the BSCS model is a specific underlying design for integrated instructional units. According to the NRC committee's report, integrated instructional units connect laboratory experience with other types of science learning activities, including reading, discussions, and lectures.

Typical (or traditional) laboratory experiences differ from the integrated instructional units in their effectiveness in attaining education goals. Research shows that typical laboratories suffer from fragmentation of goals and approaches. Research indicates that integrated instructional units are more effective than typical laboratory research for the following goals: improving mastery of subject matter, developing scientific reasoning, and cultivating interest in science. In addition, integrated instructional units appear to be effective for helping diverse groups of students progress toward these three goals.

With support from the U.S. Department of Education's Institute of Education Sciences, BSCS completed a study on the efficacy of research-based curriculum materials (i.e., integrated instructional units) with curriculum-based professional development (Taylor et al. 2015). For this research, the BSCS team pursued answers to the following questions:

- Do research-based instructional materials in multidisciplinary science units increase achievement in science for all students?
- To what extent does teacher practice mediate the effect of the curriculum materials and professional development on student achievement?
- To what extent are the effects of the materials and professional development equitable across demographic groups?

The BSCS team launched the study in 2008. It was a four-year randomized control study that tested the effectiveness of the multidisciplinary program *BSCS Science: An Inquiry Approach*. Participating schools were randomly assigned for grade 9 students to use the BSCS materials for two years, or to continue for two years with business-as-usual science instruction. The former group of teachers received seven days of curriculum implementation support. The latter group served as a comparison group, using business-as-usual curriculum materials, and participated in routine professional development. The project included 11 districts, 18 schools, 64 teachers, and 4,105 high school students. Results were aggregated and compared between the treatment and comparison conditions using annual and longitudinal measures.

The main findings included the following:

- Students participating in the BSCS science program had higher gains on standardized tests.
- By the end of 10th grade, BSCS students were four months ahead of those in the comparison group.
- The BSCS program helped teachers use more effective practices.

One of the primary ways the BSCS program attended to research on learning was to structure educational experiences using the BSCS 5E Instructional Model.

To summarize, a research-based curriculum supported by complementary professional development makes a difference in student achievement (Taylor et al., forthcoming).

RESEARCH ON THE BSCS 5E INSTRUCTIONAL MODEL

This section is based on the report *The BSCS 5E Instructional Model: Origins and Effectiveness* (Bybee et al. 2006) and more recent research on the BSCS model. Due to the relative youth of the BSCS 5E Instructional Model compared with the older SCIS Learning Cycle, there are fewer published studies that specifically compare the BSCS 5E Instructional Model with other modes of instruction. However, the findings suggest that the BSCS 5E Instructional Model, like its predecessor the SCIS Learning Cycle, is as effective—or in some cases comparatively more effective—than alternative teaching methods in helping students reach important learning outcomes in science. Several comparative studies suggest that the BSCS 5E Instructional Model is more effective than alternative approaches at helping students master science subject matter (e.g., Akar 2005; Coulson 2002). This observation relates to a later discussion on systems thinking.

In 2007, a team at BSCS conducted a laboratory-based randomized control study to examine the effectiveness of inquiry-based instruction. The team disaggregated the data by student demographic variables to examine if inquiry can provide equitable opportunities to learn. Fifty-eight students ages 14–16 were randomly assigned to one of two groups. Both groups of students were taught toward the same learning goals by the same teacher, with one group being taught from inquiry-based materials based on the BSCS 5E Instructional Model and the other from materials based on common teaching strategies as defined by national teacher survey data. Students in the inquiry-based group reached significantly higher levels of achievement than did the students experiencing common instruction. This effect was consistent across a range of learning goals—including knowledge, reasoning, and argumentation—and in time frames immediately following the instruction and four weeks later. The commonplace science instruction resulted in a detectable achievement gap by race, whereas the inquiry-based materials instruction did not.

Among other things, this study linked inquiry-based teaching to the BSCS 5E Instructional Model. Although the BSCS 5Es and inquiry are not synonymous, the instructional model is based on constructivist theories of learning and provides the structure and function for an approach to teaching that enhances student inquiry (Wilson et al. 2010).

The team measured learning outcomes for three prominent goals: scientific knowledge, scientific reasoning through application of models, and construction and critique of scientific explanations. I briefly note that these outcomes anticipate disciplinary core ideas and selected science and engineering practices described in *A Framework for K–12 Science Education* (NRC 2012) and the *Next Generation Science Standards* (*NGSS*; NGSS Lead States 2013).

The following quote is, in part, the conclusion reached by the BSCS team:

> Using scientifically based research methods required to establish causality, this study found that students receiving inquiry-based instruction reached

significantly higher levels of achievement than students experiencing common place instruction. The superior effectiveness of the inquiry-based instruction was constant across a range of learning goals (knowledge, scientific reasoning, and argumentation) and time frames (immediately following instruction and 4 weeks later). (Wilson et al. 2010, p. 291–292)

Although this study was published several years before the *NGSS*, the findings anticipate positive consequences for science and engineering practices such as Developing and Using Models, Constructing Explanations and Designing Solutions, and Engaging in Argument From Evidence.

A research team at the Center for Mathematics and Science Education at Texas A&M University conducted a longitudinal study of a fifth-grade science curriculum based on the 5E Model (Scott et al. 2014). This research answered the question, How do fifth-grade science scores change as a result of the use of a textbook that used the 5E Instructional Model?

The Region 4 Education Service Center (ESC) developed a textbook based on the 5E Model that specifically addressed the Texas state standards. Beginning in the 2005–2006 academic year, the team from Texas A&M contracted with the Region 4 ESC and a large, diverse school district that serves 13,000 students. The result was a five-year study (2005–2009) that used high-stakes achievement results of fifth graders from 14 different elementary programs. The elementary programs used the Region 4 ESC textbooks and the teachers received professional development from Region 4 curriculum specialists. The Texas Assessment of Knowledge and Skills (TAKS) for science provided the standard for measurement of student achievement. Student achievement was evaluated across four years (Scott et al. 2014).

As may be expected in a study of this scale, the team reported some technical problems. However, the team reported that the districts with achievement gaps for African American and Hispanic students were able to narrow these gaps more than they could narrow the state achievement gap during the same time period (Scott et al. 2014, p. 49).

A concluding discussion stated the following:

While the persistent achievement gap among African Americans is troubling, the fact that approximately 74% of the population of the district (Hispanic and Whites) overtook their state counterparts over the 4-year study is a powerful testimony to the efficacy of the text incorporating the 5E model in student engagement and learning in a high stakes test environment. (Scott et al. 2014, pp. 54–55)

I note the importance of professional development for the teachers implementing 5E-based curriculum materials. Although the materials in this project were thoroughly designed, there is a need for professional development so the teachers implement the program with fidelity to the innovations, especially the different phases of the 5E Model.

Fidelity to the Instructional Model

Coulson (2002) explored how varying levels of fidelity to the BSCS 5E Model affected student learning. Selected-response tests administered pre- and post-instruction were used to measure outcomes. The findings indicated that students whose teachers taught with medium or high levels of fidelity to the BSCS 5E Instructional Model experienced learning gains that were nearly double that of students whose teachers did not use the model or used it with low levels of fidelity. The impact of varying levels of fidelity to the BSCS 5E Model affected student learning. The impact of varying levels of fidelity identified here may help explain the ambiguous results of prior research by Ward and Herron (1980).

Recent Studies on Implementation Fidelity

A publication by Taylor, Van Scotter, and Coulson (2007) reported two research studies that extended and strengthened the relationship between fidelity of curriculum implementation, specifically to phases of the BSCS 5E Instructional Model and gains in student learning. The first research included case studies of four teachers field-testing a new high school science program using the BSCS 5E Instructional Model. The research identified distinctly different student learning gains for teachers who implemented the program as designed compared to teachers who implemented the program with considerable less fidelity. The learning gains were assessed using a 20-item subset of questions from the National Science Teachers Association (NSTA). The National Association of Biology Teachers (NABT) biology exam was administered at the beginning and end of the year. Developers of the curriculum being field-tested measured fidelity through classroom observation.

The second study centered on the same general question of learning gains of students whose teachers implemented a program with fidelity versus students whose teachers implemented the program with less fidelity. The study included 326 ninth-grade students and 15 teachers. Fidelity was measured using an observation protocol adapted from Horizon Research Inc., *Classroom Observation and Analytic Protocol* (HRI 2000). The rating scales quantified the extent to which teachers encouraged students to engage in metacognitive activity, communicate their understanding of concepts, and apply their understanding to new situations. Teachers who used strategies and learning sequences consistent with the 5Es at medium (basic) or high (extensive) levels had students achieve significantly higher gain scores (Taylor, Van Scotter, and Coulson 2007, p. 46).

Taylor, Van Scotter, and Coulson (2007) explain, "The observation that marked differences in achievement began at even basic use of the 5Es makes a powerful statement about the effectiveness of the instructional model" (p. 47).

The findings and implications expressed in their quotations support the effectiveness of the BSCS 5E Instructional Model and complement studies conducted at other grade levels and science disciplines (see, for example, Ates 2005; Ebrahim 2004; Lord 1997). One aside to this discussion is the essential role played by professional development so teachers can develop an understanding of curriculum materials and the instructional model that is integral to this design.

Elementary Teachers' Feedback on the 5E Model

I have found research conducted by the Australian team of Keith Skamp and Shelley Peers to be particularly insightful concerning the 5E Model (Skamp and Peers 2012). Their paper "Implementation of Science Based on the 5E Learning Model: Insights From Teacher Feedback on Trial Primary Connections Units" includes detailed analysis and recommendations for each of the phases. A bit of context will help. The Australian Academy of Science adapted a BSCS program to produce Primary Connections, a national curriculum and professional learning initiative for primary-level education. The structure and pedagogy is based on the 5E Instructional Model. Between 2005 and 2012, more than 200 teachers provided approximately 3,000 comments on 16 Primary Connections units. The approach used in this report was a qualitative analysis of teachers' comments made during field tests of the Primary Connections program. The analysis revealed many insights about the 5E Model and its use in professional development and student learning.

This study elaborated on numerous perceptions of teachers' concerns and insights about use of the 5Es as they were a part of the Primary Connections program. In general, the 5E Model had a positive impact on teachers as they incorporated the model into their planning use. Problematic feedback related to the limited understanding of the purposes of some 5E phases; for example, some teachers omitted phases.

As part of the project and research, the Australian team developed purpose statements for each phase of the 5E Model. They then used the purpose statements as the basis for teachers' review and evaluation of the 5E Model. Figure 4.1 is an adapted summary of the 5Es with purpose statements. I have included this figure because it describes well the different purposes served by each phase. The Australian team's research found that some purposes in each phase were addressed very well and had educational value for the teachers (Skamp and Peers 2012).

Figure 4.1. Purposes of the Phases in the BSCS 5E Instructional Model

ENGAGE
- Create interest and stimulate curiosity.
- Provide a meaningful context for learning.
- Raise questions for inquiry and science practices.
- Reveal students' current ideas and beliefs.

EXPLORE
- Provide experience of the phenomenon.
- Examine students' questions to test their ideas.
- Investigate questions and problems.

EXPLAIN
- Introduce concepts and practices that can be used to interpret data and construct explanations.
- Construct multimodal explanations and justify claims in terms based on evidence.
- Compare different explanations generated by students.
- Review current scientific explanations.

ELABORATE
- Use and apply concepts and explanations in new contexts.
- Reconstruct and extend explanations using different modes, such as written language, diagrammatic and graphic modes, and mathematics.

EVALUATE
- Provide an opportunity for students to review and reflect on their understanding and skills.
- Provide evidence for changes to students' understanding, beliefs, and skills.

Source: Adapted from AAS 2008.

Unanticipated Support for the 5E Model

Research headed by David Klahr and colleagues has stimulated review and discussion of the relative importance of direct instruction and discovery learning as instructional approaches to science teaching (Chen and Klahr 1999; Klahr and Nigam 2004). In the 1999 study, Chen and Klahr investigated the efficacy of different instructional approaches for an important aspect of scientific reasoning. Specifically, they intended to compare the efficacy of direct instruction versus discovery learning. They asked the question, "What is the effectiveness of different instructional strategies in children's acquisition of the domain-general strategy (Control of Variables Strategy, or CVS)?" They had children ages 7–10 design and evaluate experiments after direct instruction in CVS and without direct instruction (i.e., discovery learning). They reported that with explicit training (i.e., direct instruction), children were able to learn and transfer the basic strategy for designing uncomplicated experiments—that is, they could apply CVS (Chen and Klahr 1999).

One interesting aspect of the research conducted by Klahr and his colleagues is that their methodology actually somewhat paralleled characteristics of the BSCS 5E Instructional

Model or an integrated instructional unit. While this is evident in the articles, it is not expressed in their conclusion that direct intervention is the most effective strategy for teaching CVS. The quotations in Table 4.2 are from the methodological sections of the key articles cited in the direct instruction versus discovery learning debate. In Table 4.2, I point out the phases that parallel the BSCS 5E Instructional Model. The entire approach used by Klahr and colleagues could well be described as an integrated instructional unit that centers on students learning the key concepts of CVS.

Chen and Klahr's 1999 research article presents a well-designed study that, in my view, used an integrated instructional approach closely resembling the BSCS 5E Instructional Model. As indicated in their summary of the methodology for the intervention, Chen and Klahr used an instructional sequence that included four of the five phases in the 5E model. With an engagement phase omitted, the researchers had the students begin with an extended exploration, proceed to an explanation of CVS that included a demonstration, and then apply or elaborate CVS to these new situations, for which they used the terms *assessment*, *Transfer-1*, and *Transfer-2*.

CONCLUSION

The BSCS 5E Instructional Model is grounded in sound education theory, has a growing base of research to support its effectiveness, and has had a significant impact on science education. Although encouraging, these conclusions indicate the need to conduct research on the effectiveness of the model, including when and how it is used, and continue to refine the model based on direct research and related research on learning.

The five phases of the BSCS 5E Instructional Model are designed to facilitate the process of conceptual change. The use of this model brings coherence to different teaching strategies, provides connections among educational activities, and helps science teachers make decisions about interactions with students.

Studies of the instructional model conducted by the internal and external evaluators showed positive trends for student mastery of subject matter and interest in science. The most significant finding, however, is that there is a relationship between fidelity of use and student achievement. In other words, the BSCS 5E Instructional Model is more effective for improving student achievement when the teacher uses the model the way it is designed. Without fidelity of use, the potential positive results of the model are greatly diminished. This line of research should be continued. In addition, the research base for the BSCS 5E Instructional Model should be broadened through additional studies that compare the model's effect on goals such as mastery of subject matter, scientific reasoning, and interest and attitudes. The widespread use of the BSCS 5E Instructional Model warrants a commitment to a line of research that rivals that of the Learning Cycle at a minimum.

Table 4.2. Alignment Between Chen and Klahr's Work and the BSCS 5E Instructional Model

QUOTES FROM CHEN AND KLAHR	ALIGNMENT WITH THE BSCS MODEL	RATIONALE
"Children were presented materials in a source domain in which they performed an initial exploration."	(Possible) Engagement	The *engage* phase initiates the learning process and exposes students' current conceptions.
"Children were asked to set up experimental apparatus so as to test the possible effects of different variables."	Exploration	In the *explore* phase, students gain experience with phenomena or events.
"... included an explanation of the rationale behind controlling variables as well as examples of how to make unconfounded comparisons."	Explanation	In the *explain* phase, the teacher may give an explanation to guide students toward a deeper understanding.
"... children were presented with problems in two additional domains."	Elaboration	In the *elaborate* phase, students apply their understanding in a new situation or context.
"Part II was a pencil-and-paper post-test given two months after Part I."	Evaluation	In the *evaluate* phase, student understanding is assessed.

Source: Chen and Klahr 1999.

REFERENCES

Akar, E. 2005. Effectiveness of 5E learning cycle model on students' understanding of acid-base concepts. MS thesis, Middle East Technical University.

Ates, S. 2005. The effectiveness of learning-cycle method on teaching DC circuits to prospective female and male science teachers. *Research in Science and Technological Education* 23 (2): 213–227.

Australian Academy of Science (AAS). 2008. *Making connections: A guide for facilitators.* Canberra: Australian Academy of Science.

Bransford, J., A. Brown, and R. Cocking, eds. 1999. *How people learn: Brain, mind, experience, and school.* Washington, DC: National Academies Press.

Brooks, J. G., and M. G. Brooks. 1993. *In search of understanding: The case for constructivist classrooms.* Alexandria, VA: Association for Supervision and Curriculum Development (ASCD).

Bybee, R. W., J. A. Taylor, A. Gardner, P. Van Scotter, J. C. Powell, A. Westbrook, and N. Landes. 2006. *The BSCS 5E Instructional Model: Origins and effectiveness.* Colorado Springs, CO: Biological Sciences Curriculum Study (BSCS).

Chen, Z., and D. Klahr. 1999. All other things being equal: Acquisition and transfer of the control of variables strategy. *Child Development* 70 (5): 1098–1120.

Coulson, D. 2002. *BSCS Science: An inquiry approach—2002 evaluation findings.* Arnold, MD: PS International.

Donovan, M., and J. Bransford, eds. 2005. *How students learn: History, mathematics, and science in the classroom.* Washington, DC: National Academies Press.

Donovan, M., J. Bransford, and J. Pellegrino. 1999. *How people learn: Bridging research and practice.* Washington, DC: National Academies Press.

Driver, R., H. Asoko, J. Leach, E. Mortimer, and P. Scott. 1994. Constructing scientific knowledge in the classroom. *Educational Researcher* 23: 5–12.

Driver, R., A. Squires, P. Rushworth, and V. Wood-Robinson. 1994. *Making sense of secondary science: Research into children's ideas.* London: Routledge.

Ebrahim, A. 2004. The effects of traditional learning and a learning cycle inquiry learning strategy on students' science achievement and attitudes toward elementary science (Kuwait). *Dissertation Abstracts International* 65 (4): 1232.

Ferster, C. S., and B. F. Skinner. 1957. *Schedules of reinforcement.* New York: Appleton-Century-Crofts.

Horizon Research, Inc. (HRI). 2000. Horizon Research-2000-2001 Local systemic change classroom observation protocol. Horizon Research, Inc. *www.horizon-research.com.*

Ilg, F., and L. Ames. 1955. *Child behavior.* New York: Dell.

Klahr, D., and M. Nigam. 2004. The equivalence of learning paths in early science instruction: Effects of direct instruction and discovery learning. *Psychological Science* 15 (10): 661–667.

Lambert, L., D. Walker, D. P. Zimmerman, J. E. Cooper, M. D. Lambert, M. E. Gardner, and M. Szabo. 1995. *The constructivist leader.* New York: Teachers College Press.

Lawson, A. E., M. Abraham, and J. Renner. 1989. *A theory of instruction: Using the learning cycle to teach science concepts and thinking skills.* NARST Monograph, Number One.

Lord, T. R. 1997. A comparison between traditional and constructivist teaching in college biology. *Innovative Higher Education* 21 (3): 1127–1147.

Matthews, M. 1992. Constructivism and the empirist legacy. In *Relevant research: Scope, sequence, and coordination of secondary school science, vol. II,* ed. M. K. Pearsall, 183–196. Washington, DC: NSTA Press.

National Research Council (NRC). 2000. *Inquiry and the National Science Education Standards: A guide for teaching and learning.* Washington, DC: National Academies Press.

National Research Council (NRC). 2006. *America's lab report: Investigations in high school science.* Washington, DC: National Academies Press.

National Research Council (NRC). 2012. *A framework for K–12 science education: Practices, crosscutting concepts, and core ideas.* Washington, DC: National Academies Press.

NGSS Lead States. 2013. *Next Generation Science Standards: For states, by states.* Washington, DC: National Academies Press. *www.nextgenscience.org/next-generation-science-standards*

Piaget, J. 1975. From noise to order: The psychological development of knowledge and phenocopy in biology. *Urban Review* 8 (3): 209.

Piaget, J. 1976. Piaget's theory. In *Piaget and his school: A reader in developmental psychology,* ed. C. Zwingmann, B. Inhelder, and H. H. Chipman, pp. 11–23. New York: Springer-Verlag.

Piaget, J., and B. Inhelder. 1969. *The psychology of the child.* New York: Basic Books.

Posner, G. J., K. A. Strike, P. W. Hewson, and W. A. Gertzog. 1982. Accommodation of a scientific conception: Toward a theory of conceptual change. *Science Education* 66 (2): 211–227.

Scott, T. P., C. Schroeder, H. Tolson, T. Huang, and O. M. Williams. 2014. A longitudinal study of a 5th grade science curriculum based on the 5E Model. *Science Educator* 23 (1): 49–55.

Skamp, K., and S. Peers. 2012. Implementation of science based on the 5E Learning Model: Insights from teacher feedback on trial Primary Connections units. Paper presented at the Australian Science Education Research Association Conference, Queensland, Australia.

Taylor, J., S. Getty, S. Kowalski, C. Wilson, J. Carlson, and P. Van Scotter. Forthcoming. An efficacy trial of research-based curriculum materials with curriculum-based professional development. *American Educational Research Journal.*

Taylor, J., P. Van Scotter, and D. Coulson. 2007. Bridging research on learning and student achievement: The role of instructional materials. *Science Educator* 16 (2): 44–50.

Vygotsky, L. S. 1962. *Thought and language.* Cambridge, MA: MIT Press.

Ward, C., and J. Herron. 1980. Helping students understand formal chemical concepts. *Journal of Research in Science Teaching* 17 (5): 387–400.

Wilson, C. D., J. A. Taylor, S. M. Kowalski, and J. Carlson. 2010. The relative effects and equity of inquiry-based and commonplace science teaching on students' knowledge, reasoning, and argumentation. *Journal of Research in Science Teaching* 47 (3): 276–301.

Using the 5E Model to Implement the *Next Generation Science Standards*

Thisis chapter provides recommendations for translating standards into instructional materials that are usable for those with the real task of teaching. The discussion provides an affirmative answer to the question, How can the BSCS 5E Instructional Model be used to implement the *Next Generation Science Standards* (*NGSS*)? I recommend beginning with a review of *A Framework for K–12 Science Education: Practices, Crosscutting Concepts, and Core Ideas* (NRC 2012) and becoming familiar with the *Next Generation Science Standards: For States, by States* (NGSS Lead States 2013). *A Reader's Guide to the* Next Generation Science Standards (Pratt 2013) would also provide helpful background and resources.

The BSCS 5E Instructional Model can be used as the basis for instructional materials that align with the aims of *NGSS*. In fact, the instructional model proves to be quite helpful as an organizer for the instructional sequences required to accommodate the three dimensions of performance expectations in *NGSS*. I have described this process in significant detail in *Translating the* NGSS *for Classroom Instruction* (Bybee 2013) and recommend that book for those deeply involved in the task of developing or adapting instructional materials based on *NGSS*. This chapter draws on insights I gained during my work on both the *National Science Education Standards* (NRC 1996) and the *NGSS* (NGSS Lead States 2013); the process of writing the book on translating the *NGSS* for classroom instruction required developing examples of classroom instruction that may be of interest.

ENGAGING IN *NGSS* AND CLASSROOM INSTRUCTION

How would you apply the 5E Model to *NGSS*? What would you consider as central to the process? Think about how you would answer these questions in the contexts of your classroom and your students.

EXPLORING *NGSS*

The Anatomy of a Standard

Let's begin by briefly reviewing a standard. Figure 5.1 (p. 64) is a standard for first-grade life science. I selected this example because it is simple and presents elements that clarify the anatomy of a standard.

One can view the standard as the box at the top of the framework. This is one perspective for a standard. Due to states' requirements, what is defined as a standard is ambiguous in *NGSS*. I have found it most helpful to focus on the performance expectations as they define the competencies that serve as the learning outcomes for instruction and assessments. Notice the standard is headed by Heredity: Inheritance and Variation of Traits. The subhead is "Students who demonstrate understanding can …" This is followed by a statement identified with the number and letters "1-LS-3." Statement 1 describes a performance expectation. In the case of this standard, the performance expectation is, "Make observations to construct an evidence-based account that young plants and animals are like, but not exactly like, their parents."

Very important, performance expectations specify a set of learning outcomes. That is, they illustrate the *competencies* students should develop as a result of classroom instruction. At this point, it is important to note that the performance expectations are specifications for assessments with implications for curriculum and instruction, but they are neither instructional units or teaching lessons, nor actual classroom tests.

Performance expectations embody three essential dimensions: science and engineering practices, disciplinary core ideas, and crosscutting concepts. The three columns beneath the performance expectation are statements from *A Framework for K–12 Science Education* (NRC 2012) and provide detailed *content* for the three dimensions in performance expectations.

Figure 5.1. Heredity: Inheritance and Variation of Traits Standard From *NGSS*

1-LS3 Heredity: Inheritance and Variation of Traits

1-LS3   Heredity: Inheritance and Variation of Traits
Students who demonstrate understanding can:
1-LS3-1. Make observations to construct an evidence-based account that young plants and animals are like, but not exactly like, their parents. [Clarification Statement: Examples of patterns could include features plants or animals share. Examples of observations could include leaves from the same kind of plant are the same shape but can differ in size; and, a particular breed of dog looks like its parents but is not exactly the same.] [Assessment Boundary: Assessment does not include inheritance or animals that undergo metamorphosis or hybrids.]
The performance expectations above were developed using the following elements from the NRC document *A Framework for K-12 Science Education*:

Science and Engineering Practices	**Disciplinary Core Ideas**	**Crosscutting Concepts**
Constructing Explanations and Designing Solutions Constructing explanations and designing solutions in K–2 builds on prior experiences and progresses to the use of evidence and ideas in constructing evidence-based accounts of natural phenomena and designing solutions. • Make observations (firsthand or from media) to construct an evidence-based account for natural phenomena. (1-LS3-1)	**LS3.A: Inheritance of Traits** • Young animals are very much, but not exactly, like their parents. Plants also are very much, but not exactly, like their parents. (1-LS3-1) **LS3.B: Variation of Traits** • Individuals of the same kind of plant or animal are recognizable as similar but can also vary in many ways. (1-LS3-1)	**Patterns** • Patterns in the natural world can be observed, used to describe phenomena, and used as evidence. (1-LS3-1)

Connections to other DCIs in this grade-level: will be available on or before April 26, 2013.
Articulation of DCIs across grade-levels: will be available on or before April 26, 2013.
Common Core State Standards Connections: will be available on or before April 26, 2013. ELA/Literacy – Mathematics –

*The performance expectations marked with an asterisk integrate traditional science content with engineering through a Practice or Disciplinary Core Idea.

Source: NGSS Lead States 2013.

To further understand standards, we can dissect the performance expectation. Look at performance expectation 1 in Figure 5.1: Make observations to construct an evidence-based account that young plants and animals are like, but not exactly like, their parents. *Making observations to construct an explanation* is the practice. Look in the foundation box on the left for Constructing Explanations and Designing Solutions and find the bullet statement "Make observations (firsthand or from media) to construct an evidence-based account for natural phenomena." Details for the disciplinary core idea are in the center of the foundation column under Inheritance of Traits and Variation of Traits. Finally, the crosscutting concept, Patterns, is described in the right column. All three descriptions are keyed to the performance expectation as indicated by 1-LS3-1.

The box beneath the three content columns provides *connections* to *Common Core State Standards* for English language arts and mathematics and the articulation of this standard to other topics at the grade level and across grade levels.

With this brief introduction to *NGSS* and the competencies, we can move to the translation from the standard—the performance expectation—to the instructional model.

EXPLAINING A PROCESS FOR APPLYING THE 5E MODEL TO *NGSS*

Thinking Beyond a Lesson to an Integrated Instructional Sequence

Expanding conceptions about instruction from a daily lesson to an integrated instructional sequence will be helpful when translating *NGSS* to classroom instruction. Here is a metaphor that clarifies my suggestion: Life sciences recognize the cell as the basic unit of life. There also are levels at which cells are organized—tissues, organs, organ systems, and organisms. While the lesson remains the basic unit of instruction, in translating the *NGSS* to classroom instruction, it is essential to expand one's perception of science teaching to other levels of organization such as a coherent, integrated sequence of instructional activities. By analogy, think about organ systems, not just cells. Although the idea of instructional units has a long history, a recent analysis of research on laboratory experience in school science programs (NRC 2006) presents a perspective of integrated instructional units that connect laboratory experience with other types of learning activities, including reading, discussions, and lectures. The BSCS 5E Instructional Model is a helpful way to think about an integrated instructional unit (see Figure 5.2, p. 66). The 5E Model provides the general framework for the translation of *NGSS* to classroom instruction.

Figure 5.2. Integrated Instructional Sequence

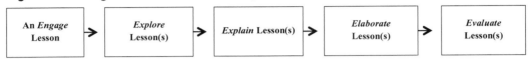

The next sections of this chapter present several insights and lessons learned as a result of translating *NGSS* performance expectations for elementary, middle, and high school classrooms.

The process of actually translating standards to classroom practices was, for me, a very insightful experience. To say the least, the process is more complex than I realized. But my familiarity with the 5E Model was a great help in figuring out how to design classroom instruction based on *NGSS*.

Identify a Coherent Set of Performance Expectations

The examples in Figure 5.2 focused on a single performance expectation. I did this for simplicity and clarity. Here, I move to a discussion of a coherent set of performance expectations (i.e., a cluster or bundle) and recommend not identifying single performance expectations with single lessons. The process of translating performance expectations is much more efficient if one considers a coherent set of performance expectations that make scientific and educational sense.

Begin by examining a standard with the aim of identifying a cluster of performance expectations that form a topic of study that may be appropriate for a two- to three-week unit. Components of the disciplinary core ideas, major themes, topics, and conceptual ideas represent ways to identify a coherent set of performance expectations. Topics common to science programs may help identify a theme for an instructional sequence. The primary recommendation is to move beyond thinking about each performance expectation as a lesson; try to identify a theme that would be the basis for a unit of study that incorporates several performance expectations. This is a reasonable way to begin thinking about translating standards to school programs and classroom practices. In the prior example, Figure 5.2, the unit might be Heredity and Variation of Traits.

With this recommendation stated, in some cases you may find that one performance expectation does require a single lesson sequence or that all of the performance expectations in a standard can be accommodated in a single unit of instruction.

Distinguish Between Learning Outcomes and Instructional Strategies

The scientific and engineering practices may be viewed *both* as teaching strategies and learning outcomes. Of particular note is the realization that the scientific and engineering practices as learning outcomes also represent both knowledge and abilities for the instructional sequence. As learning outcomes, one wants students to develop the abilities and knowledge that these practices are basic to science and engineering.

As you begin applying the instructional model, bear in mind that students can, in using instructional strategies, actively ask questions, define problems, develop models, carry out investigations, analyze data, use mathematics, construct explanations, engage in arguments, and communicate information and understand that each of these science and engineering practices is a learning outcome. In applying the 5E Model, you should distinguish between the teaching strategies and learning outcomes—for the student. Using the practices as teaching strategies does not necessarily mean students will learn the practices.

Consider How to Integrate Three Learning Outcomes—Science and Engineering Practices, Crosscutting Concepts, and Disciplinary Core Ideas

Recognize that a performance expectation describes a set of three learning outcomes and criteria for assessments. This recommendation begins by considering—thinking about, reflecting on, and pondering—how the three dimensions might be integrated in a carefully designed sequence of activities. Taken together, the learning experiences should contribute to students' development of the scientific or engineering practices, crosscutting concepts, and disciplinary core ideas.

Beginning with *A Framework for K–12 Science Education* (NRC 2012), continuing to the *Next Generation Science Standards* (NGSS Lead States 2013) and now translating those standards to using the 5E Instructional Model, one of the most significant challenges has been that of integration. It is easy to recommend (or even require) that the three dimensions be integrated, but much more complex to actually realize this integration in classroom instruction. The teams developing standards solved the problem in the statements of performance expectations. Now the challenge moves to curriculum and instruction.

Several fundamentals of integrating a science curriculum may help. These lessons are paraphrased from a study (BSCS 2000) and article that colleagues and I completed (Van Scotter, Bybee, and Dougherty 2000). First, do not worry about what you call the integrated instructional sequence; instead, consider what students will learn. Second, regardless of what you integrate, coherence must be the essential quality of the instruction and assessment. Third, the fundamental goal of any science curriculum, including an integrated one, should be to increase students' understanding of science concepts (both core and crosscutting), and science and engineering practices and their ability to apply those concepts and practices. Begin with an understanding that concepts and practices will be integrated across an instructional sequence, then proceed by identifying activities, investigations, or engineering problems that may be used as the basis for instructional sequence.

Apply the BSCS 5E Instructional Model

Use the 5E Model as the basis for a curriculum unit. While lessons serve as daily activities, design the sequence of lessons using a variety of experiences (e.g., web searches, group investigations, readings, discussions, computer simulations, videos, direct instruction) that contribute to the learning outcomes described in the performance expectations.

Here are the four principles of instructional design that contribute to attaining learning goals as stated in *NGSS*. First, instructional materials are designed with clear performance expectations in mind. Second, learning experiences are thoughtfully sequenced into the flow of classroom instruction. Third, the learning experiences are designed to integrate learning of science concepts (i.e., both disciplinary core ideas and crosscutting concepts) with learning about the practices of science and engineering. Finally, students have opportunities for ongoing reflection, discussion, discourse, and argumentation.

Use Backward Design

Understanding by Design (Wiggins and McTighe 2005) describes a process that will enhance science teachers' abilities to attain higher levels of student learning. The process is called *backward design*. Conceptually, the process is simple. Begin by identifying your desired learning outcomes—for example, the performance expectations from *NGSS*. Then determine what would count as acceptable evidence of student learning. You should formulate strategies that set forth what counts as evidence of learning for the instructional sequence. This should be followed by actually designing assessments that will provide the evidence that students have learned the competencies described in the performance expectations. Then, and only then, begin developing the activities that will provide students opportunities to learn the concepts and practices described in the three dimensions of the performance expectations.

The dimensions of scientific and engineering practices, crosscutting concepts, and disciplinary core ideas as described in the *A Framework for K–12 Science Education* (NRC 2012) and the performance expectations and foundation boxes in the *NGSS* (NGSS Lead States 2013) describe learning outcomes. They are the basis for using backward design for the development or adaptation of curriculum and instruction. Performance expectations also are the basis for assessments. Simply stated, the performance expectation can and should be the starting point for backward design.

The BSCS 5E Instruction Model and the *NGSS* provide practical ways to apply the backward design process. Let us say you identified a unit and performance expectations for Life Cycles of Organisms. One would describe concepts and practices to determine the acceptable evidence of learning. For instance, students would need to use evidence to construct an explanation that clarifies life cycles of plants and animals, identify aspects of the cycle (e.g., being born, growing to adulthood, reproducing, and dying), and describe the patterns of different plants and animals. You might expect students to recognize that offspring closely resemble their parents and that some characteristics are inherited from parents while others result from interactions with the environment. Using the BSCS 5E Instructional Model, one could first design an *evaluate* activity, such as growing Fast Plants under different environmental conditions and designing a rubric with the aforementioned criteria. Then, one would proceed to design the *engage, explore, explain*, and *elaborate* experiences. As necessary, the process would be iterative between the *evaluate* phase and other activities as the development process progresses. Figure 5.3 presents the backward design process and the 5E Instructional Model.

Figure 5.3. Backward Design Process and the 5E Instructional Model

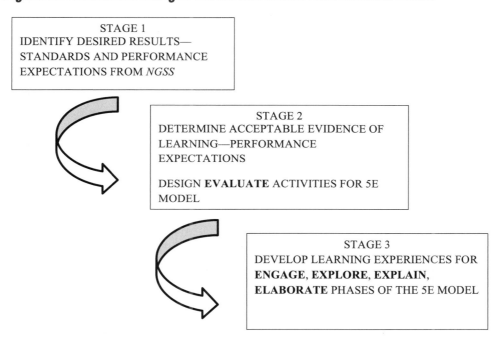

STAGE 1
IDENTIFY DESIRED RESULTS—
STANDARDS AND PERFORMANCE
EXPECTATIONS FROM *NGSS*

STAGE 2
DETERMINE ACCEPTABLE EVIDENCE OF
LEARNING—PERFORMANCE
EXPECTATIONS

DESIGN **EVALUATE** ACTIVITIES FOR 5E
MODEL

STAGE 3
DEVELOP LEARNING EXPERIENCES FOR
**ENGAGE, EXPLORE, EXPLAIN,
ELABORATE** PHASES OF THE 5E MODEL

Source: Adapted from Wiggins and McTighe 2005.

Remember to Include Engineering and the Nature of Science

Standards in the *NGSS* include the performance expectations. The standards describe the competencies or learning goals and are best placed in the first stage when applying backward design. The performance expectations and the content described in foundation boxes beneath the performance expectations represent acceptable evidence of learning and a second stage in the application of backward design. One caution should be noted: Sometimes use of the scientific and engineering practices combined with the crosscutting concepts and disciplinary core ideas is interpreted as a learning activity that would be included in Stage 3. The caution is to include the activity in Stage 2—as a learning outcome.

Stage 3 involves development or adaptation of activities that will help students attain the learning outcomes.

In *NGSS*, some performance expectations emphasizing engineering and the nature of science are included. It is important to identify these (*Note:* They are identified in the scientific and engineering practices and crosscutting concepts columns of the foundation boxes). Because they are described as practices or crosscutting concepts, they should be integrated along with the disciplinary core ideas. Their recognition calls for a different emphasis in the instructional sequence.

Recognize Opportunities to Emphasize Different Learning Outcomes

As you begin adapting activities or developing materials, be aware of opportunities to emphasize science or engineering practices, crosscutting concepts, and disciplinary core ideas within the 5E instructional sequence. This is an issue of recognizing when one of the three dimensions can be explicitly or directly emphasized—move it from the background (i.e., not directly or explicitly emphasized) of instruction to the foreground (i.e., clearly and directly emphasized). To understand my use of foreground and background, think of a picture. Usually there is something (e.g., a person) in the foreground and other features in the background. The foreground is what the photographer emphasized, and the background provides context (e.g., the location of the picture). You can apply the idea of foreground and background to curriculum and instruction. For curriculum materials of instructional practices, what is emphasized (foreground) and what is the context (background)? Furthermore, as one progresses through the 5E instructional sequence, different aspects of performance expectations can be in the foreground or background. This curricular emphasis is indicated in Table 5.1 by the words *foreground* and *background* in the framework's cells.

I must clarify this recommendation. Although the three dimensions are integrated, the intention is that students learn the concepts and abilities of all three. The probability of students learning a practice, for example, that is in the background and used as an instructional strategy is less likely than using the same practice for instruction *and* making it explicit and directly letting students know that this is a scientific or engineering practice.

Completing a framework such as the one displayed in Table 5.1 provides an analysis of the three dimensions and can serve as feedback about the balance and emphasis of the three dimensions within the 5E instructional sequence and, subsequently, the need for greater or lesser emphasis on particular dimensions.

Table 5.1. A Framework for Applying the BSCS 5E Instructional Model to *NGSS* Performance Expectations

INSTRUCTIONAL SEQUENCE	SCIENCE AND ENGINEERING PRACTICES	DISCIPLINARY CORE IDEAS	CROSSCUTTING CONCEPTS
Engage	Foreground Background	Foreground Background	Foreground Background
Explore	Foreground Background	Foreground Background	Foreground Background
Explain	Foreground Background	Foreground Background	Foreground Background
Elaborate	Foreground Background	Foreground Background	Foreground Background
Evaluate	Foreground Background	Foreground Background	Foreground Background

EXPANDING YOUR UNDERSTANDING OF *NGSS* AND THE 5E MODEL

In this section, you actually extend your understanding by translating a performance expectation from the *NGSS* to a sequence of classroom instruction. For simplicity and convenience, you can begin with the first-grade life science performance expectation you explored in a prior section. That standard is displayed in Figure 5.1 (see p. 64).

Using this performance expectation and related information in the foundation boxes and connections, design an instructional sequence using the 5E Model. You should complete the framework in Table 5.2 (p. 72) by describing what the teacher does and what the students do.

I selected this *NGSS* standard because it presented less complexity from a practice, core idea, and crosscutting concept point of view. It also is the case that you had already explored the standard and gained some understanding of the performance expectation, foundational content, and connections.

Now that you have completed this process, you may wish to identify a set of performance expectations for a discipline and grade level of relevance to you. This activity would give a second elaboration, and one that should be more complex.

Table 5.2. Applying the BSCS 5E Instructional Model to *NGSS* Standards

THE BSCS 5E INSTRUCTIONAL MODEL	WHAT THE TEACHER DOES	WHAT THE STUDENT DOES
Engagement: This phase of the instructional model initiates the learning task. The activity should make connections between past and present learning experiences, surface any misconceptions, and anticipate activities that reveal students' thinking on the learning outcomes of current activities. The student should become mentally engaged in the concepts, practices, or skills to be explored		
Exploration: This phase of the teaching model provides students with a common base of experiences within which they identify and develop current concepts, practices, and skills. During this phase, students may use cooperative learning to explore their environment or manipulate materials.		
Explanation: This phase of the instructional model focuses students' attention on a particular aspect of their engagement and exploration experiences and provides opportunities for them to verbalize their conceptual understanding or demonstrate their skills or behaviors. This phase also provides opportunities for teachers to introduce a formal label or definition for a concept, practice, skill, or behavior.		
Elaboration: This phase of the teaching model challenges and extends students' conceptual understanding and allows further opportunity for students to practice desired skills and behaviors. Cooperative learning is appropriate for this stage. Through new experiences, the students develop deeper and broader understanding, more information, and adequate skills.		
Evaluation: This phase of the teaching model encourages students to assess their understanding and abilities and provides opportunities for teachers to evaluate student progress toward achieving the performance expectation.		

EVALUATING YOUR INSTRUCTIONAL SEQUENCE

You can use a modification of criteria for adapting instructional materials for an evaluation of your understanding of the 5E model and *NGSS*. Table 5.3 describes the criteria, questions for evaluation, and your analysis.

Table 5.3. Evaluating Your Application of the BSCS 5E Instructional Model to *NGSS*

CRITERIA	QUESTIONS FOR THE ANALYSIS	YOUR ANALYSIS
• Identification of scientific and engineering practices • Crosscutting concepts • Disciplinary core and component ideas	• Do topics of the instructional sequence match the three dimensions of *NGSS*? • Are standards explicitly represented in the sequence?	
• Explicit connections among practices, crosscutting concepts, and disciplinary core and component ideas	• Do activities include the practices, crosscutting concepts, and disciplinary core ideas of the standards? • Do activities include all the component ideas? • Are connections made with other topics, concepts, and practices?	
• Time and opportunities to learn	• Does instruction include several experiences on a dimension? • Do students experience concepts before vocabulary is introduced? • Do students apply concepts and practices in different contexts?	
• Appropriate and varied instruction	• Are different methods of instruction used? • Are students engaged in activities that emphasize all three dimensions?	
• Appropriate and varied assessment	• Do you first identify what students know and do? • Are assessment strategies consistent with the performance expectations? • Are assessments comprehensive, coherent, and focused on the integration of core and component ideas, crosscutting concepts, and science and engineering practices?	
• Potential connections to *Common Core State Standards* for English language arts and mathematics	• Where does the instructional sequence present opportunities to make connections to the *Common Core State Standards?*	

CONCLUSION

Based on lessons I learned in translating *NGSS* to classroom instruction, this chapter provides helpful insights for those who have the task of applying the BSCS 5E Instructional Model. Additionally, the chapter modeled the 5E instructional sequence for addressing a performance expectation.

REFERENCES

Biological Sciences Curriculum Study (BSCS). 2000. *Making sense of integrated science: A guide for high schools.* Colorado Springs, CO: BSCS.

Bybee, R. 2013. *Translating the* NGSS *for classroom instruction.* Arlington, VA: NSTA Press.

National Research Council (NRC). 1996. *National science education standards.* Washington, DC: National Academies Press.

National Research Council (NRC). 2006. *America's lab report: Investigations in high school science.* Washington, DC: National Academies Press.

National Research Council (NRC). 2012. *A framework for K–12 science education: Practices, crosscutting concepts, and core ideas.* Washington, DC: National Academies Press.

NGSS Lead States. 2013. *Next Generation Science Standards: For states, by states.* Washington, DC: National Academies Press. *www.nextgenscience.org/next-generation-science-standards*

Pratt, H. 2013. *The NSTA reader's guide to the* Next Generation Science Standards. Arlington, VA: NSTA Press.

Van Scotter, P., R. W. Bybee, and M. J. Dougherty. 2000. Fundamentals of integrated science. *The Science Teacher* 67 (6): 25–28.

Wiggins, G., and J. McTighe. 2005. *Understanding by design.* Alexandria, VA: Association for Supervision and Curriculum Development (ASCD).

Applying the 5E Model to STEM Education

The acronym STEM (science, technology, engineering, and mathematics) emerged in the education community in the 1990s. Since then, use of STEM has continually expanded to encompass almost anything related to these areas. In education, STEM is quite popular and has a varied, if ambiguous, meaning. In the context of this book, one could easily ask the question, Can the BSCS 5E Instructional Model be applied to STEM education? Given the diversity in meanings, the reasonable answer is that it depends. I will reframe the question: If one is concerned about a STEM curriculum and classroom instruction, can the 5E Model be applied? Here the answer is yes. In fact, the positive response to the question seems particularly appropriate given the inclusion of engineering design in the *Next Generation Science Standards* (*NGSS*; NGSS Lead States 2013) and mathematics in the *Common Core State Standards* (*CCSS*; NGAC and CCSSO 2010).

A SOCIETAL PERSPECTIVE FOR STEM EDUCATION

The 20th century was a period of significant scientific advances and technological innovations, both of which contributed to dramatic social progress. As the nation's economy advanced, the requirements for skilled workers increased, especially the need for intellectual skills, including those often associated with STEM fields.

By 21st-century standards, the intellectual skills required in the early 20th century were low. With time, the nation's policy makers and educators realized the economic value of creative ideas and efficient means for the production and delivery of goods and services. As the 20th century progressed, the number of individual jobs requiring manual labor and routine cognitive skills steadily decreased, while the jobs requiring intellectual abilities such as adapting ideas and solving nonroutine problems increased. In short, work became more analytical and technical. By the 20th century's end, entry-level requirements for the workforce increased to levels beyond a high school education. Taking this general observation to a more specific level, one would have to note the combined role of science, technology, engineering, and mathematics as a driving force of economic change and the steady shift in requirements for entry into the workforce, especially in developed countries. The changes just described suggest a fundamental place for science, technology, engineering, and mathematics in our economy, and by extension in our education programs.

A CLASSROOM EXAMPLE OF THE BSCS 5E INSTRUCTIONAL MODEL AND A STEM TOPIC

The following discussion is based on an adaptation of a released report from the 2006 Program for International Student Assessment (PISA). The title of the PISA unit was The Greenhouse Effect: Fact or Fiction? The unit can be reviewed in *Green at Fifteen: How Fifteen-Year-Olds Perform in Environmental Science and Geoscience in PISA 2006* (OECD 2009).

The original unit, cited above, was presented as an assessment that exemplified the PISA approach and the levels of proficiency for different items. In the following discussion, I have adapted this unit by adding a classroom and teaching context and arranging the activities to align with the 5E Instructional Model.

Engaging Students in a STEM Issue

The class is middle school, grade 8. This is the first day of the unit. On day 1 of the unit, the teacher, Mr. Kennedy, engages the students with a teachable moment. Mr. Kennedy enters the classroom and tells the students, "I have a puzzling situation. The other day I mentioned global warming and one student became very interested. That student, Andy, became so engaged that he went on the web and did a search of the topic and found two graphs that he thought showed a possible relationship between the average temperature of the Earth's atmosphere and the carbon dioxide (CO_2) emissions on the Earth. At this point in the lesson, Mr. Kennedy showed the students the graphs that Andy discovered (Figure 6.1).

Figure 6.1. Graphs Showing Possible Relationship Between the Average Temperature of Earth's Atmosphere and CO_2 Emissions on Earth

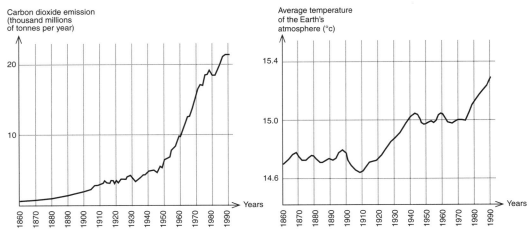

Source: OECD. 2009. *Green at fifteen?: How 15-year-olds perform in environmental science and geoscience in PISA 2006.* PISA, OECD Publishing. *http://dx.doi.org/10.1787/9789264063600-en.* Reprinted with permission.

Mr. Kennedy explained that Andy concluded from these two graphs that it is certain that the increase in the average temperature of the Earth's atmosphere is due to the increase in the carbon dioxide emission.

At this point, Mr. Kennedy divided the class into pairs and asked the groups to discuss what they concluded from the two graphs and Andy's position. Mr. Kennedy provided each group with the graphs. In this *engage* phase, the teacher created a teachable moment by suggesting that he was puzzled and wanted the students to review the two graphs that initiated Andy's interest and formed the basis for his conclusion. He did not define the terms *global warming* or *greenhouse effect*. Furthermore, he did not clarify the details of the two graphs.

The small-group discussion provided an opportunity for the teacher to assess the students' knowledge of the greenhouse effect, their understanding of the graphs, and their reasoning that connected the graphs to Andy's conclusion.

Exploring the STEM Issue

On day 2, Mr. Kennedy further explained the problem: Another student, Jean, disagreed with Andy's conclusion. She compared the two graphs and said that some parts of the graphs did not support his conclusion.

Mr. Kennedy asked the small groups to identify what evidence might have been the basis for Jean's claim and her disagreement with Andy's conclusion. The teacher asked Jean to give an example of the part of the graphs that did not support Andy's conclusion.

Different groups of students presented the following observations:

- In the period 1900–1910, CO_2 emissions increased and temperatures decreased.
- In the period 1980–1983, CO_2 emissions decreased and temperatures increased.
- In the 1800s, the temperature stayed about the same and CO_2 emissions increased.
- Between 1950 and 1980, CO_2 emissions increased but the temperature did not.
- Between 1940 and 1975, the temperature stayed about the same, but CO_2 emissions increased significantly.

In some other groups, the students made incomplete observations. For example, some groups only indicated dates and other groups did not indicate a period of time, only a date with an acceptable example. Finally, some groups identified the difference between the two graphs but did not refer to specific periods that supported Andy's conclusion. Mr. Kennedy asked the students to indicate the data from the graphs and their reasoning—that is, how and why the data support Andy's conclusion. As the teacher listened to the student discussions, he observed the following ideas. Some groups referred to the increase of both temperature and CO_2 emissions. Some even used the term *correlation* to describe the similarity in graphs. Some examples included these statements:

- Both graphs indicate increases—one of CO_2 emissions, the other of temperature.
- In 1910, both graphs increased.
- The higher the CO_2 emissions, the higher the temperature.

Explaining Components of a STEM Issue

The class begins with a representative of the student groups explaining what they observed about the two graphs and the possible conclusions by Andy and Jean.

After listening to the students' explanations, Mr. Kennedy told the class that he would like them to listen while he introduced some scientific ideas that would help them with the activity. He proceeded to discuss the following ideas:

- Living things need energy to survive. The energy that sustains life on Earth comes from the Sun, which radiates energy into space. A tiny proportion of this energy reaches the Earth.
- The Earth's atmosphere acts like a protective blanket over the surface of our planet, preventing the variations in temperature that would exist in an airless world.
- Most of the radiated energy coming from the Sun passes through the Earth's atmosphere. The Earth absorbs some of this energy, and some is reflected back from the Earth's surface. Part of the reflected energy is absorbed by the atmosphere.
- As a result of absorbed energy, the average temperature above the Earth's surface is higher than it would be if there were no atmosphere. The Earth's atmosphere has the same effect as a greenhouse, hence the term *greenhouse effect*.
- Scientists report that the greenhouse effect has become more pronounced during the 20th century.
- Scientific evidence indicates that the average temperature of the Earth's atmosphere has increased. Increased carbon dioxide emission is often cited as the main source of the temperature rise in the 20th century.

After this introduction, Mr. Kennedy asks for students' questions. The students ask for more information on the greenhouse effect, how scientists know the atmospheric temperature is changing, why some light energy gets through the Earth's atmosphere but does not get out, and how much evidence scientists have about changes in the Earth's atmosphere.

Mr. Kennedy gives further explanations and examples in response to these questions. In this phase of the instructional sequence, the teacher first asked for students' explanations as indicators of what they had learned. The second part of this lesson was a direct introduction of concepts related to the greenhouse effect and the evidence supporting the concept. Finally, there was time for questions from the students.

Elaborating on STEM Knowledge and Practices

Mr. Kennedy begins the next class by telling the students that Andy persists in his conclusion that the average rise in temperature of the Earth's atmosphere is caused by the increase in carbon dioxide emissions. But Jean thinks his conclusion is premature. She says, "Before accepting this conclusion, you must be sure that other factors that could influence the greenhouse effect are constant."

The students are then assigned the task of preparing two statements. One statement should support Andy's conclusion and the other should support Jean's position. The students should use the original graphs, their own web searches, and other information in the preparation of their statements. The students will hand in the written statements and make a presentation to the class about their conclusions.

Evaluating STEM Knowledge and Practices

To evaluate students' learning, Mr. Kennedy designed the following rubric (see Table 6.1).

Table 6.1. Evaluation Rubric for Greenhouse Lesson

LEARNING OUTCOMES	EVALUATION	
Knowledge of greenhouse effect	Formative	Summative
Analyzing and interpreting data		
Using mathematics and computational thinking		
Constructing explanations		
Engaging in argument from evidence		
Obtaining, evaluating, and communicating information		
Systems thinking		

RECOMMENDATIONS

Applying the BSCS 5E Instructional Model to STEM education is possible; the example used in this chapter demonstrates that possibility. This section presents several recommendations that will help you use the 5E Model for STEM education.

Identify a Context for Your Unit of Instruction

I have found it particularly difficult to begin with a discipline (e.g., science) and then try to build an integrated instructional sequence that includes other disciplines (e.g., technology, engineering, mathematics). My recommendation is to begin with a local, regional, or global context that has personal meaning for your students. Although there may be others, contexts that lend themselves to STEM include the following:

- Health maintenance and disease prevention
- Energy efficiency
- Environmental quality
- Natural hazards
- Natural resource use

The example in the prior section used global climate change as the context.

Decide on Your Approach to Integration

A step beyond maintaining separate STEM disciplines requires the consideration and a decision to advance STEM education by integrating the disciplines. This decision can be made at the state level, but in the approach suggested here, the decision is best made at the district or school level.

Several approaches to curriculum integration have been published. I recommend reviewing the following resources: *Designs for Science Literacy* (AAAS 2001); *Meeting Standards Through Integrated Curriculum* (Prak and Burns 2004); *Making Sense of Integrated Science: A Guide for High Schools* (BSCS 2000); and *Interdisciplinary Curriculum: Design and Implementation* (Jacobs 1989).

In addition, the National Academy of Engineering and the National Research Council (NRC) released *STEM Integration in K–12 Education* (NRC 2014). I also note that some of this discussion is adapted from my book *The Case for STEM Education: Challenges and Opportunities* (Bybee 2013).

Different perspectives of STEM education can be described. Here are several variations to consider for the integration of STEM:

- *Coordinate:* Two subjects taught in separate courses are coordinated so content in one subject synchronizes with what is needed in another subject. For example, students in mathematics learn algebraic functions when they need that knowledge in engineering.
- *Complement:* While teaching the main content of one subject, the content of another subject is introduced to complement the primary subject. For example,

while designing an energy-efficient car in a technology class, science concepts of frictional resistance (drag), loss of kinetic energy, and mass are introduced to improve the car's design and efficiency.

- *Correlate:* Two subjects with similar themes, content, or processes are taught so students understand the similarities and differences. For example, you could teach scientific practices and engineering design in separate science and technology courses.

- *Connections:* The teachers use one discipline to connect other disciplines. For example, they could use technology as the connection between science and mathematics.

- *Combine:* This approach combines two or more STEM disciplines using projects, themes, procedures, or other organizing foci. For example, one could establish a new course on science and technology that uses student projects to show the relationship between science and technology.

Because of the dominance of the traditional disciplines in state, district, and school standards, curricula, and assessments, you likely will need to provide a rationale with supporting recommendations for integrating the STEM disciplines. This is especially the case when you move beyond integration through coordination, complements, correlation, or connections. Combining subjects or designing courses that transcend the separate STEM disciplines will require elaborate and detailed justifications.

There are a few arguments for curricular integration. First, the situations of life and living are all integrated. The decisions that citizens face are not nicely contained within disciplines such as science or mathematics. Life situations typically require the knowledge, abilities, and skills of multiple disciplines. Second, individuals learn best when the context within which they are learning has personal meaning—that is, learning is enhanced when it is related to something people recognize or know, or in which they have a personal interest. Third, there is an efficiency that comes with combining the knowledge and skills of different disciplines, and there is limited time in school days and years. If lessons, courses, and school programs can attain learning outcomes of both content and processes of different disciplines such as engineering and mathematics, that benefits both teachers and students.

Table 6.2. Applying the BSCS 5E Instructional Model to a STEM Topic

THE BSCS 5E INSTRUCTIONAL MODEL	WHAT THE TEACHER DOES	WHAT THE STUDENT DOES
Engagement: This phase of the instructional model initiates the learning task. The activity should make connections between past and present learning experiences, surface any misconceptions, and anticipate students' thinking on the learning outcomes of current activities. The student should become mentally engaged in the concept, practices, or skill to be explored.		
Exploration: This phase of the teaching model provides students with a common base of experiences within which they identify and develop current concepts, practices, and skills. During this phase, students may use cooperative learning to explore their environment or manipulate materials.		
Explanation: This phase of the instructional model focuses students' attention on a particular aspect of their engagement and exploration experiences and provides opportunities for them to verbalize their conceptual understanding or demonstrate their skills or behaviors. This phase also provides opportunities for teachers to introduce a formal label or definition for a concept, practice, skill, or behavior.		
Elaboration: This phase of the teaching model challenges and extends students' conceptual understanding and allows further opportunity for students to practice desired skills and behaviors. Cooperative learning is appropriate for this stage. Through new experiences, the students develop deeper and broader understanding, more information, and adequate skills.		
Evaluation: This phase of the teaching model encourages students to assess their understanding and abilities and provides opportunities for teachers to evaluate students' progress toward achieving the performance expectation.		

Developing an Instructional Sequence for a STEM Topic

Table 6.2 on page 82 briefly describes criteria for each phase of the 5E model. After you have identified an appropriate STEM topic, use the table to sketch an instructional sequence. I strongly recommend using backward design as it was presented in Chapter 5.

CONCLUSION

The acronym STEM is widely used in education. Although STEM has caught the interest of policy makers and many educators, the meaning remains elusive. In the 21st century, citizens need to have essential knowledge and skills associated with science, technology, engineering, and mathematics. This chapter provided a rationale for STEM education and directed the reader's attention to STEM as it may be applied in the context of curriculum and instruction. Thus, an application of the 5E Model gains specific meaning for classroom teachers.

REFERENCES

American Association for the Advancement of Science (AAAS). 2001. *Designs for science literacy.* Washington, DC: AAAS.

Biological Sciences Curriculum Study (BSCS). 2000. *Making sense of integrated science: A guide for high schools.* Colorado Springs, CO: BSCS.

Bybee, R. 2013. *The case for STEM education: Challenges and opportunities.* Arlington, VA: NSTA Press.

Jacobs, H. 1989. *Interdisciplinary curriculum: Design and implementation.* Alexandria, VA: Association for Supervision and Curriculum Development (ASCD).

National Governors Association Center for Best Practices and Council of Chief State School Officers (NGAC and CCSSO). 2010. *Common core state standards.* Washington, DC: NGAC and CCSSO.

National Research Council (NRC). 2014. *STEM integration in K–12 education: Status, prospects, and agenda for research.* Washington, DC: National Academies Press.

NGSS Lead States. 2013. *Next Generation Science Standards: For states, by states.* Washington, DC: National Academies Press. *www.nextgenscience.org/next-generation-science-standards.*

Office for Economic Cooperation and Development (OECD). 2009. *Green at fifteen: How 15-year-olds perform in environmental science and geoscience in PISA 2006.* Paris, France: OECD.

Prak, S., and R. Burns. 2004. *Meeting standards through integrated curriculum.* Alexandria, VA: Association for Supervision and Curriculum Development (ASCD).

Extending the 5E Model to 21st-Century Skills

I n late 2008, the economy experienced the greatest series of crises since the Great Depression. This economic downturn continued into 2009 and continues experiencing a slow recovery. Not surprisingly, the public's attention has centered on jobs and basic needs. In the midst of discussions about bailouts and stimulus packages, there also was discussion of a deep technical workforce and skills required in the 21st century. Although the nation's economic problems are larger and more complex than workforce skills, recovery and retraining of the retired, underemployed, and unemployed will require a new generation with knowledge, attitudes, and abilities generally needed for the 21st century. Predictably, some discussion turned to education's role in response to the economic problems.

The discussion of education and workforce skills is not new. For example, in 1983, the Task Force on Education for Economic Growth at the Education Commission of the States prepared *Action for Excellence: A Comprehensive Plan to Improve Our Nation's Schools.* In 1984, the National Academies published the report *High Schools and the Changing Workplace* (NRC 1984), and in 1991, the U.S. Department of Labor released *What Work Requires of Schools: A SCANS Report for America 2000.*

The recent emphasis on skills includes *Teaching the New Basic Skills* (Murnane and Levy 1996) and more than 20 reports expressing the need to address concerns about an unprepared or underprepared workforce. Among the most notable of these reports is *Rising Above the Gathering Storm: Energizing and Employing America for a Brighter Economic Future* (NRC 2005 and 2007).

Now educators have the challenge of clarifying the skills and moving from broad statements of purpose to more specific discussions of education practice. In this chapter, I address potential connections between development of 21st-century skills and the BSCS 5E Instructional Model. This chapter draws on a report for the National Institutes of Health, Office of Science Education, which was prepared by BSCS (Bybee et al. 2006).

21ST-CENTURY WORKFORCE SKILLS AND THE BSCS 5E INSTRUCTIONAL MODEL

In 2007, the National Academies (U.S.) held workshops that identified five broad skills that accommodated a range of jobs, from low-skill, low-wage service to high-wage, high-skill professional work. Individuals can develop these broad skills within classrooms and programs, as well as in other settings (NRC 2008, 2010; Levy and Murnane 2004).

Research indicates that individuals learn and apply broad 21st-century skills within the context of specific bodies of knowledge (NRC 2008, 2010; Levy and Murnane 2004). At work, development of these skills is intertwined with development of technical content knowledge. Similarly, in classrooms students may develop cognitive skills while engaged in the study of specific STEM-related social or global situations. The following discussion presents five skill sets that are important for the 21st century. Those skill sets include adaptability, complex communications, nonroutine problem solving, self-management, and systems thinking. These skills are summarized from the NRC report *Exploring the Intersection of Science Education and 21st Century Skills* (2010) and a more detailed report *Education for Life and Work: Developing Transferable Knowledge and Skills in the 21st Century* (NRC 2012).

Before turning to the specific issue of 21st-century skills and the potential of the BSCS 5E Instructional Model to contribute to the development of those skills, this section clarifies several points.

As prior sections have noted, fidelity to the structure and sequence of the 5E Model results in greater student learning. This is not to say that teachers cannot adapt or modify teaching strategies with nuances that may enhance learning, but the modifications should be made with knowledge and understanding of the design of the curriculum material and the instructional model. The decisions must be informed by an understanding of the learning theory underlying the instructional model (e.g., Bransford, Brown, and Cocking 2000).

A review of the BSCS 5E Instructional Model did not find any cases where the model was used explicitly for development of the 21st-century skills. Here, I would note the importance of establishing the 21st-century skills as explicit learning outcomes. In this case, it seems educational policies may be leading curriculum programs and instructional practices designed for teachers to implement 21st-century skills (Gewertz 2008).

Finally, I suggest that one or several of the skills should be emphasized in the context of other content. One would not develop a 5E sequence just for a particular skill, such as adaptability. Rather, within an exploration, for example, the teacher would point out the need to adapt by responding to different ideas, learning new approaches, listening and responding to other members of the group, and modifying procedures for an investigation.

Adaptability

This skill includes the ability and willingness to cope with uncertain, new, and rapidly changing conditions on the job, including responding effectively to emergencies or crisis situations and learning new tasks, technologies, and procedures. Adaptability also includes handling work stress; adapting to different personalities, communication styles, and cultures; and adapting physically to various indoor or outdoor work environments (Houston 2007; Pulakos et al. 2000).

Implications for the 5E Model especially center on the *explore* and *elaborate* phases. These phases often involve group work and the challenges of dealing with new situations, learning new procedures, and solving problems. In general, these skills would be in the background of activities, and the teacher's role would be to make students aware of the need to learn new procedures and adjust to different personalities. The instructional strategy within the 5E phase would likely be that of coaching for adaptability as the need arises in different contexts.

Complex Communications and Social Skills

These skills include processing and interpreting both verbal and nonverbal information from others before responding appropriately. A skilled communicator selects key pieces of a complex idea to express in words, sounds, and images to build shared understanding (Levy and Murnane 2004). Skilled communications include facilitating positive outcomes with customers, subordinates, and superiors through social perceptiveness, persuasion, negotiation, instructing, and service orientation (Peterson et al. 1999).

This skill clearly complements goals described in the *Common Core State Standards* for literacy and the connections between the performance expectations in the *Next Generation Science Standards* (*NGSS*) and the *Common Core*. The opportunities to develop these skills also center on activity- and project-based experiences, ones with a clear orientation of *NGSS* and STEM. Students should engage in opportunities to collect data and present their findings using graphs, charts, or other means. Arguments based on evidence would be fundamental to developing the skills. Relative to the 5E Model, the complex communication would apply to activities in the *explore*, *elaborate*, and *evaluate* phases. They would be especially important for activities that require developing evidence-based explanations and argumentation.

I refer the reader to the problem about greenhouse gases introduced in Chapter 6. Mr. Kennedy's assignment in the *elaborate* phase would require a high level of complex communication skills for the students. Although the primary discussion of the classroom example in Chapter 6 focused on STEM, one can imagine the 21st-century skills in the background of activities.

Nonroutine Problem Solving

These skills can be characterized by an individual using expert thinking to examine a broad span of information, recognizing patterns, and narrowing the information to diagnose a problem. Moving beyond diagnosis to a solution requires knowledge of how the information is linked conceptually and involves metacognition—the ability to reflect on whether a problem-solving strategy is working and to switch to another strategy if the current strategy is not working (Levy and Murnane 2004). Nonroutine problem solving includes creating new and innovative solutions, integrating seemingly unrelated information, and entertaining possibilities (Houston 2007).

Certainly, creativity and innovation are part of problem solving, and project-based activities and evidence-oriented activities would be central to these skills and abilities of nonroutine problem solving. As students engage in scientific inquiry, they have the afore-mentioned opportunities. The studies by Taylor, Van Scotter, and Coulson (2007); Boddy, Watson, and Aubusson (2003); and Wilson et al. (2010) suggest positive support for the BSCS 5E Instructional Model and contribute to a link between scientific reasoning and problem solving. Addressing the engineering design elements in *NGSS* has direct implications for nonroutine problem-solving abilities and all phases of the 5E model, especially those that involve problem-based activities, such as the *explore*, *elaborate*, and *evaluate* phases.

Self-Management and Self-Development

Here, the emphasis is on the personal skills needed to work remotely and in virtual teams, to work autonomously, and to be self-motivating and self-monitoring. One aspect of self-management is the willingness and ability to acquire new information and skills related to work (Houston 2007). Underlying these skills, one assumes interest in, motivation to learn about, and positive attitudes toward a domain of study or work. Research supporting the role of the BSCS 5E Instructional Model in developing interest can be found in studies by Von Secker (2002), Akar (2005), and Tinnin (2000).

In general, the types of activities supported by *NGSS* and STEM provide opportunities for students to work alone, identify areas to study, investigate, and acquire new knowledge.

Systems Thinking

This important skill includes understanding how an entire system works and how an action, change, or malfunction in one part of the system affects other components of the system—thus, adopting a "big picture" perspective on work (Houston 2007). This skill includes decision making, systems analysis, and systems evaluation as well as abstract reasoning about how the different elements of a work process interact (Peterson et al. 1999; Meadows 2008).

Understanding how a system works and how changes in components may affect the entire system is, in my view, critical knowledge for 21st-century citizens. In addition, the skills of analyzing a system, recognizing the role of subsystems, and understanding the structure and functions of systems are essential abilities.

This ability requires mastery of knowledge about systems and the application of that knowledge to practical laboratory work and contextual situations. The majority of studies based on the BSCS 5E Instructional Model support the efficacy of the model to enhance students' mastery of subject matter (see, for example, Bybee et al. 2006; Coulson 2002;

Taylor, Van Scotter, and Coulson 2007; Akar 2005; and Wilson et al. 2010). Additional support for systems thinking is from findings by Boddy, Watson, and Aubusson (2003), who reported increased higher-order thinking by students after a unit of work based on the BSCS 5E Instructional Model.

Systems thinking represents an exception to my earlier statement that 21st-century skills are background content. I believe that systems thinking represents essential knowledge and ability and, as such, should be in the foreground of an instructional sequence based on the 5E Model.

These 21st-century skills reveal a mixture of cognitive abilities, social skills, personal motivation, conceptual knowledge, and problem-solving competencies. Although diverse, this knowledge and these skills and abilities can be developed in school programs that include scientific inquiry, technological innovation, and mathematical computation. Individual materials and classroom instruction designed to address contemporary standards such as the *NGSS* and *Common Core State Standards* for English language arts and math will contribute to development of 21st-century skills for all students. The BSCS 5E Instructional Model will contribute to achieving this aim.

Views of Science Teachers

A member poll by the National Science Teachers Association (NSTA) reported that educators thought 21st-century skills were vital for students but challenging to impart. Not surprisingly, teachers emphasized core subject knowledge and learning and innovation skills. But many teachers struggle with how to teach knowledge and abilities. Furthermore, they indicated difficulties when it came to measuring the effectiveness of instruction relative to 21st-century skills (NSTA 2014).

The NSTA survey included the five skills described in this chapter. Results of the survey indicate that a majority (60%) of educators do not consider 21st-century skills when planning lessons. Of course, the good news is that 40% *do* consider 21st-century skills when planning. Which skills take precedence in planning science lessons? In decreasing order of perceived importance, these skills are nonroutine problem solving (50%), complex communication and social skills (41%), self-management and self-development (36%), systems thinking (20%), and adaptability (18%). The next sections describe ways to address the challenge of implementing 21st-century skills.

Relationship Between the BSCS 5E Model and 21st-Century Skills

A study by BSCS staff called "The Relative Effects and Equity of Inquiry-Based and Commonplace Science Teaching on Students' Knowledge, Reasoning and Argumentation: A Randomized Control Trial" (Wilson et al. 2010) examined the effectiveness of inquiry-based curriculum materials and teaching (e.g., the BSCS 5E Instructional Model). Fifty-eight students ages 14–16 were randomly assigned to one of two groups. Both groups were taught toward the same goals by the same teacher, with one group being taught from materials organized around the BSCS 5E Instructional Model and the other from commonplace teaching strategies as defined by national teacher survey data. Students in the group where the teacher used the BSCS model reached significantly higher levels of achievement compared to the other group. The effect was consistent for the range of learning goals—knowledge, scientific reasoning, and argumentation. The finding held for testing immediately following instruction and four weeks later. This study lends support to the 21st-century goals of nonroutine problem solving (i.e., scientific reasoning), complex communication (i.e., argumentation), and systems thinking (i.e., knowledge).

Curriculum Goals

Addressing one of the first challenges would be including 21st-century skills in the curriculum goals. I realize the goals of 21st-century skills are relatively new, but they can be addressed in the context of many activities, investigations, and projects. So, one recommendation is to explicitly state 21st-century skills as program goals. One would begin with the five goals described in this chapter.

Instructional Sequence

Assuming the 5E Model is used as the instructional sequence, one can review each phase, looking for situations that lend themselves to 21st-century skills. When do students work in small groups? When are students asked to communicate? What problem are students trying to solve? When are students required to manage their time? What is the system under study? Rather than thinking in terms of including a lesson on complex communication, for example, identify instructional phases that present opportunities to *briefly* identify and emphasize this skill in the context of ongoing work.

Contextual Opportunities

Just as 21st-century skills are applied and emphasized in workplace contexts, they can be in classroom work as well. As opportunities present themselves in various phases of the 5E Model, use that moment to highlight a skill or ability. I would recommend mentioning the skills and abilities in the *explain* phase of the 5E Model so students realize your goals and understand that the skills and abilities will be included on an assessment.

Assess the Skills and Abilities

As you design both formative and summative assessments, be sure to include the 21st-century skills emphasized in the instructional sequence and unit.

CONCLUSION

Can the BSCS 5E Instructional Model be used to develop 21st-century workforce skills? Yes, it can.

The BSCS 5E Instruction Model and other such models do hold promise for teaching 21st-century skills. This said, it also must be noted that although the development of skills and abilities has been noted as an educational goal for school programs, very little emphasis has been placed on these goals.

Activity-based school programs that incorporate the BSCS 5E Instructional Model have the potential to develop 21st-century skills. Among the major challenges associated with this assertion are providing model curriculum materials that exemplify the goals, supporting the curriculum model with professional development, changing teachers' perceptions about explicitly teaching to develop skills and abilities, and encouraging fidelity to instructional models designed to help students attain 21st-century skills.

REFERENCES

Akar, E. 2005. Effectiveness of 5E learning cycle model on students' understanding of acid-base concepts. MS thesis, Middle East Technical University.

Boddy, N., K. Watson, and P. Aubusson. 2003. A trial of the five Es: A referent model for constructivist teaching and learning. *Research in Science Education* 33 (1): 27–42.

Bransford, J., A. Brown, and R. Cocking, eds. 2000. *How people learn: Brain, mind, experience, and school.* Expanded ed. Washington, DC: National Academies Press.

Bybee, R. W., J. A. Taylor, A. Gardner, P. Van Scotter, J. C. Powell, A. Westbrook, and N. Landes. 2006. *The BSCS 5E instructional model: Origins and effectiveness.* Colorado Springs, CO: Biological Sciences Curriculum Study (BSCS).

Coulson, D. 2002. *BSCS Science: An inquiry approach—2002 evaluation findings.* Arnold, MD: PS International.

Gewertz, C. 2008. States press ahead on "21st-century skills." *Education Week,* October 15, 2008: 21–23.

Houston, J. 2007. Future skill demands: From a corporate consultant perspective. Presentation at the Workshop on Research Evidence Related to Future Skill Demands, National Academies of Science, Washington, DC. *www7.nationalacademies.org/cfe/future_skill_demands_presentations.html.*

Levy, F., and R. Murnane. 2004. *The new division of labor: How computers are creating the next job market.* Princeton, NJ: Princeton University Press.

Meadows, D. 2008. *Thinking in systems.* White River Junction, VT: Chelsea Green Publishing.

Murnane, R., and F. Levy. 1996. *Teaching the new basic skills: Principles for educating children to thrive in a changing economy.* New York: Free Press.

National Research Council (NRC). 1984. *High schools and the changing workplace.* Washington, DC: National Academies Press.

National Research Council (NRC). 2005, 2007. *Rising above the gathering storm: Energizing and employing America for a brighter economic future.* Washington, DC: National Academies Press.

National Research Council (NRC). 2008. *Research on future skill demands: Workshop summary.* Margaret Hilton, Rapporteur. Washington, DC: National Academies Press.

National Research Council (NRC). 2010. *Exploring the intersection of science education and 21st century skills.* Washington, DC: National Academies Press.

National Research Council (NRC). 2012. *Education for life and work: Developing transferable knowledge and skills in the 21st century.* Washington, DC: National Academies Press.

National Science Teachers Association (NSTA). 2014. NSTA members poll: 21st century skills vital for students, challenging to impart, say educators. *NSTA Reports* 25 (7): 8–9.

Peterson, N., M. Mumford, W. Borman, P. Jeanneret, and E. Fleishman. 1999. *An occupational information system for the 21st century: The development of O'NET.* Washington, DC: American Psychological Association (APA).

Pulakos, E., S. Arad, M. Donovon, and K. Plamondon. 2000. Adaptability in the workplace: Development of a taxonomy of adaptive performance. *Journal of Applied Psychology* 85 (4): 612–624.

Task Force on Education for Economic Growth. 1983. *Action for excellence: A comprehensive plan to improve our nation's schools.* Denver, CO: Education Commission of the States.

Taylor, J., P. Van Scotter, and D. Coulson. 2007. Bridging research on learning and student achievement: The role of instructional materials. *Science Educator* 16 (2): 44–50.

Tinnin, R. 2000. The effectiveness of a long-term professional development program on teachers' self-efficacy, attitudes, skills, and knowledge using a thematic learning approach. *Dissertation Abstracts International* 6 (11): 43–45.

U. S. Department of Labor. 1991. *What work requires of schools: A SCANS report for America 2000.* Washington, DC: U.S. Government Printing Office.

Von Secker, C. 2002. Effects of inquiry-based teacher practices on science excellence and equity. *The Journal of Educational Research* 95: 151–160.

Wilson, C. D., J. A. Taylor, S. M. Kowalski, and J. Carlson. 2010. The relative effects and equity of inquiry-based and commonplace science teaching on students' knowledge, reasoning, and argumentation: A randomized control trial. *Journal of Research in Science Teaching* 47 (3): 276–301.

EVALUATE

Assessing Understanding and Use

I Was Just Wondering ... : An Evaluation

Across the years, I have seen and been asked many questions about the BSCS 5E Instructional Model. Often the questions begin with, "I was just wondering." On most occasions, the questioner is either already using or considering use of the BSCS 5E Instructional Model. This chapter serves as an evaluation by addressing some of the questions and issues raised by curriculum developers and classroom teachers.

The first section of this chapter presents a self-evaluation of your understanding and use of the BSCS 5E Instructional Model. I pose a series of questions and ask that you reflect on what you understand of the 5E Model and answer the questions. In the chapter's next section, I provide answers to the questions.

A SELF-EVALUATION

Briefly answer the following questions:

- How will I know if I created a teachable moment?

- What is the appropriate use of the instructional model—a lesson, an activity, a unit, a school program? Please explain your response.

- Can the order of phases be shifted?

- Can a phase be omitted?

- Can a phase or phases be added?

- Can phases be repeated?

- Should evaluation be continuous?

- What if I need to explain an idea before (or after) the *explain* phase?

- Where has the BSCS 5E Instructional Model been implemented?

- What if I wanted to design my own instructional model?

- Can the BSCS 5E Instructional Model be used for the *NGSS* (or other innovations such as STEM education, 21st-century skills, or the *Common Core State Standards*)?

MY RESPONSES

How Will I Know If I Have Created a Teachable Moment?

Recently, it has been common for speakers to respond to a question by complimenting the person asking the question, stating, "Good question." Well, this really *is* a good question. It deserves an honest answer. You may know you have created a teachable moment by outward, visible, and verbal responses. Students will look puzzled, perplexed, or confounded. Depending on the experience that captures their interest, they may respond by saying things such as, "How did that happen?" or "I wondered about that?" or "That really confused me because I thought something else would happen."

It also is the case that some students may not overtly demonstrate their state of puzzlement. They may be thinking, trying to resolve their confusion, or formulating an explanation without outward behavior or statements.

The reality of classrooms is that there may be 25 (or more) students and they will not all be engaged at the same time. Use of the 5E Model does increase the probability of the classroom teacher creating a teachable moment for more students more of the time.

What Is the Appropriate Use of the Instructional Model?

More specifically, should the instructional model be the basis for one lesson? A unit of study? An entire program? My experience suggests that the optimal use of the BSCS 5E Instructional Model is a unit of two or three weeks, where each phase is used as the basis for one or more lessons (with the exception of the *engage* phase, which may be less than a lesson). In this recommendation, I do not equate a phase with a daily lesson. The *engage* phase may be less than a daily lesson and other phases may be more than a daily lesson. I also assume some repetition or cycling of lessons within a phase; for example, there might be two lessons in the *explore* phase and three lessons in the *elaborate* phase.

Using the 5E Model as the basis for a single lesson decreases the effectiveness of the individual phases due to shortening the time and opportunities for challenging and restructuring of concepts and abilities—for learning. On the other hand, using the phases for an entire program greatly increases the time and experience of the individual phases, so the perspective for the phase loses its effectiveness. This occurs, for example, because students have too much exploration or multiple explanations that may reduce the emphasis on core concepts for the discipline.

Can a Phase Be Omitted?

My recommendation: Do not omit a phase. Earlier research on the Science Curriculum Improvement Study (SCIS) Learning Cycle found a decreased effectiveness when phases were omitted or their position shifted (Lawson, Abraham, and Renner 1989). From a contemporary understanding of how students learn, there is integrity to each phase and the sum of the phases as originally designed (Taylor, Van Scotter, and Coulson 2007). This question is often based on prior ideas about instruction that would omit the *engage* or *explore* phases and go immediately to the *explain* phase. Alternatively, some suggest omitting the *elaborate* phase. My point here centers on the experiences required for transfer of learning combined with the application of knowledge in a new but related context.

Can the Order of Phases Be Shifted?

My response to this question is similar to the prior one on omitting a phase: Do not shift the order. What would one shift? Move the *explain* phase prior to *explore*? The original sequence was designed to enhance students' learning and has been supported by research (see, for example, NRC 1999a, 1999b; Bybee et al. 2006; Wilson et al. 2010). There also is earlier research on the learning cycle that specifically investigated the question about changing the sequence (Renner, Abraham, and Bernie 1988; Marek and Cavallo 1997). That research indicated reduced effectiveness of the model when the sequence was changed, so I do not recommend shifting phases.

Can a Phase or Phases Be Added?

A colleague added two phases by splitting the *engage* phase into *elicit* and *engage* phases and adding an *extend* after *evaluate* to underscore the importance of knowledge transfer (Eisenkraft 2003). In principle, I do not have a problem with adding a phase (or two) if the justification is grounded in research on learning, which was the case for the modification just described.

Can Phases Be Repeated?

Yes, it is sometimes necessary for teachers to repeat a phase. This change should be based on the curriculum developer's or teacher's judgment relative to students' need for time and experiences to learn a concept or develop ability. To be clear, an example of repeating a phase would be *engage, explore, explore*—not necessarily changing the order by, for example, placing an *explore* before the *evaluate*.

Should Evaluation Be Continuous?

Yes. Effective teachers continuously evaluate their students' understanding, so I respond positively to this question. In the instructional model, the *evaluate* phase is intended as a summative assessment conducted at the end of the instructional sequence or unit. Certainly, some evaluation ought to be informal and continuous. Continuous assessment might indicate the need to add another *exploration*, for example, before students are ready to move into the *explanation* phase. But there also is need for a formal evaluation at the end of the unit. As described in earlier discussions, evaluation in the 5E Model should take the form of an activity in which the students first describe their understanding and then complete a formal assessment that provides the teacher with evidence about the students' level of understanding and abilities.

What If I Need to Explain an Idea Before (or After) the *Explain* Phase?

This may be necessary, as some ideas are prerequisites to students' understanding the primary concepts of a unit. Teachers will have to make a judgment about the priority and prerequisite nature of the concepts. One should maintain an emphasis on the primary or major concepts and abilities of the unit and avoid digressing with less-than-essential explanations. Ask questions such as, "Is this explanation essential?" or "Does it fall into categories of nice to know or interesting, but not essential?"

Where Has the BSCS 5E Instructional Model Been Implemented?

The model has been widely implemented in education. This widespread use falls into three primary categories of use: (1) policy documents that frame larger pieces of work, such as curriculum frameworks, assessment guidelines, or course outlines; (2) curriculum materials of various lengths and sizes; and (3) adaptations for teacher professional development, informal education settings, and disciplines other than science. A simple internet search using a popular search engine such as Google reveals the wide and varied applications of the BSCS 5E Instructional Model. A search in 2006 (Bybee et al. 2006) showed the following range of uses:

- More than 235,000 lesson plans developed and implemented using the BSCS 5E Instructional Model

- More than 97,000 posted and discrete examples of universities using the 5E Model in their course syllabi

- More than 73,000 examples of curriculum materials developed using the 5E Model

- More than 131,000 posted and discrete examples of teacher education programs or resources that use the 5E Model

- At least three states that strongly endorsed the 5E Model, including Texas, Connecticut, and Maryland

It is reasonable to assume the numbers have only increased since 2006. So, there is an extraordinary range of teachers and students who have been exposed to the model.

What If I Wanted to Design My Own Instructional Model?

There are design principles that account for the efficacy of the BSCS 5E Model, and those principles are applicable to a variety of curriculum goals. Figure 8.1 lists design principles for an instructional model that could be used by curriculum developers and classroom teachers. These principles are adapted from an earlier publication (Bybee 1997).

Figure 8.1. Design Principles for an Instructional Model

1. The model should have between three and five phases that represent an integrated instructional sequence.
2. The model should be based on contemporary research on student learning and development.
3. The model must help the learner integrate new skills and abilities with prior skills and abilities.
4. The model must allow for social interactions (student-student as well as student-teacher interactions).
5. The model must be generic and applicable to a wide range of classroom contexts and activities.
6. The model must be manageable for teachers with classrooms of 25 or more students.
7. The model must be understandable to teachers and students.
8. The model must accommodate and incorporate a variety of teaching strategies, including laboratories, educational technologies, reading, writing, and individual student work.

A word of caution: Discussions about curriculum development and instruction often emphasize topics and content for a course of study. While a topic might be part of an instructional sequence, an instructional model emphasizes research on how students learn *concepts*, not topics.

Can the BSCS 5E Instructional Model Be Used for the *NGSS* (or Other Innovations Such as STEM Education, 21st-Century Skills, or the *Common Core State Standards*)?

Yes, I have actually found the 5E Model helps solve the challenge of incorporating the multiple dimensions of *NGSS* in the classroom. The phases of instruction certainly can include activities that afford opportunities for students to experience the science and engineering practices, disciplinary core ideas, and crosscutting concepts. In *Translating the* NGSS *for Classroom Instruction* (Bybee 2013), I used the 5E Model for examples of the integration of multiple dimensions of *NGSS*.

In addition, the 5E Model can be used for STEM education, in particular for the design of curriculum materials, and as an approach that provides opportunities to develop 21st-century skills.

Relative to the *Common Core State Standards* (*CCSS*), using the 5E Model in contexts such as *NGSS*, STEM, or 21st-century skills does present situations where connections to *CCSS* can be made.

A FINAL EVALUATION

I wonder how well you understand the BSCS 5E Instructional Model. Figure 8.2 presents brief narratives of different segments of an instructional sequence. As presented in Figure 8.2, the excerpts are *not* in a sequence that describes the 5E Model. Review the brief narratives and place them in a 5E sequence. Use the letters and number for the sequence you think works best. Place your evaluation in the boxes. You can check your answers at the end of this chapter.

☐ ☐ ☐ ☐ ☐ ☐

Figure 8.2. An Instructional Sequence: Final Evaluation

5	Ms. Smith asked the students to explain what they knew about systems and how they would describe a system. Students' answers were varied, but she listened and accepted their explanations. Ms. Smith then told the class about systems and gave several examples that students might group: school system and transportation system.
E	Ms. Smith stated the lesson's purpose—to develop an understanding of systems and subsystem concepts. She continued with a brief definition: A system is an interconnected set of elements that is organized in a way that achieves a goal. She asked the students to identify the three key aspects of the definition (i.e., elements, interconnections, and function or purpose) and give examples. After some struggle, the students came up with answers. Ms. Smith complimented the students, telling them they did a good job and that she wanted them to identify at least one system at home that evening.
I	Ms. Smith asked each student to use the definition of a system and write a description of a system. Students' descriptions included cars, dishwashers, computers, televisions, and toasters. For each example, the teacher asked the student to clarify the elements, interconnections, and purpose of the system.
T	Ms. Smith divided students into pairs and distributed a battery, bulb, and wire to each pair. She challenged the students to light the bulb using only the battery and wire. The students tried various ways of connecting the battery and bulb. After several minutes, one group, then another, and another lit their bulbs. As students made connections and got the bulbs to light, they expressed satisfaction, even glee. With eyes wide open, they smiled and showed each other and Ms. Smith their accomplishments.
B	Ms. Smith asked different groups to tell what they did to get the bulb to light. As the students described their processes, Ms. Smith had them clarify the different elements of the challenge, the connections that worked and didn't work, and how they knew when they had accomplished the task. As the student descriptions continued, Ms. Smith did not introduce science concepts such as energy and electrical circuits.
M	The students were presented with different systems. The systems included a sports team, a highway, a tree, and their city. In teams of two, they were asked to select one of the systems and give a brief oral description of how the system works in terms of the elements, interconnections, and purpose. After 20 minutes, the teams gave their oral reports.

CONCLUSION

The statement that heads this chapter—"I was just wondering"—is an example of a teachable moment for teachers interested in creating teachable moments. I am encouraged when teachers understand enough about the 5E Model to ask questions and seek a deeper understanding of the model and its use in classrooms.

Answers to the Final Evaluation

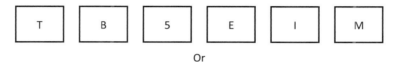

| T | B | 5 | E | I | M |

Or

The BSCS 5 E Instructional Model

REFERENCES

Bybee, R. 1997. *Achieving scientific literacy.* Portsmouth, NH: Heinemann.

Bybee, R. 2013. *Translating the* NGSS *for classroom instruction.* Arlington, VA: NSTA Press.

Bybee, R. W., J. A. Taylor, A. Gardner, P. Van Scotter. J. C. Powell, A. Westbrook, and N. Landes. 2006. *The BSCS 5E instructional model: Origins and effectiveness.* Colorado Springs, CO: Biological Sciences Curriculum Study (BSCS).

Eisenkraft, A. 2003. Expanding the 5E model. *The Science Teacher* 70 (6): 56–59.

Lawson, A., M. Abraham, and J. Renner. 1989. *A theory of instruction.* NARST Monograph, Number One. Cincinnati, OH: National Association for Research in Science Teaching.

Marek, E., and A. Cavallo. 1997. *The learning cycle: Elementary school science and beyond.* Portsmouth, NH: Heinemann.

National Research Council (NRC). 1999a. *How people learn: Brain, mind, experience, and school.* Washington, DC: National Academies Press.

National Research Council (NRC). 1999b. *How people learn: Bridging research and practice.* Washington, DC: National Academies Press.

Renner, J. W., M. R. Abraham, and H. H. Bernie. 1988. The necessity of each phase of the learning cycle in teaching high school physics. *Journal of Research in Science Teaching* 25 (1): 39–58.

Taylor, J., P. Van Scotter, and D. Coulson. 2007. Bridging research on learning and student achievement: The role of instructional materials. *Science Educator* 16 (2): 44–50.

Wilson, C. D., J. A. Taylor, S. M. Kowalski, and J. Carlson. 2010. The relative effects and equity of inquiry-based and commonplace science teaching on students' knowledge, reasoning, and argumentation. *Journal of Research in Science Teaching* 47 (3): 276–301.

Implementing the BSCS 5E Instructional Model in Your Classroom: A Final Evaluation

H ere is a final evaluation. Ask, "How can I begin using the 5E Model in my classroom?" If you ask this question, I am very encouraged. Your interest and understanding must be high. You are more than engaged! Even more important, taking steps to answer the question will be important. One of the best ways you can begin using the model is with a series of lessons you already teach. It may be the case that you begin with one lesson and expand that lesson into a series of lessons using the 5E Model as the basis. You should note that the prior chapters on the *Next Generation Science Standards* (*NGSS*), STEM, and 21st-century skills all demonstrate a series of lessons, not a single lesson based on the 5E Model. I also encourage you to use backward design as you revise your lessons for the 5E Model. So, we begin the final evaluation with a brief review of backward design and then progress to the development of an integrated instructional sequence using the BSCS 5E Instructional Model.

USING CURRENT LESSONS TO IMPLEMENT THE 5E MODEL

Begin With Backward Design

Understanding by Design (Wiggins and McTighe 2005) describes a process that will enhance teachers' abilities to attain higher levels of student learning. The process is called *backward design*. Conceptually the process is simple. Begin by identifying your desired learning outcomes—for example, the aims of your instructional sequence. Next, determine what would count as acceptable evidence of student learning. After determining what would count as evidence, design assessments that will provide the evidence that students have learned the outcomes described in the aims. Then, and only then, begin developing and sequencing the activities that will provide students the opportunities to learn the concepts and skills described in the aims.

Figure 9.1 (p. 104) describes the process of backward design as you will use it in this chapter.

Figure 9.1. Backward Design and the BSCS 5E Instructional Model

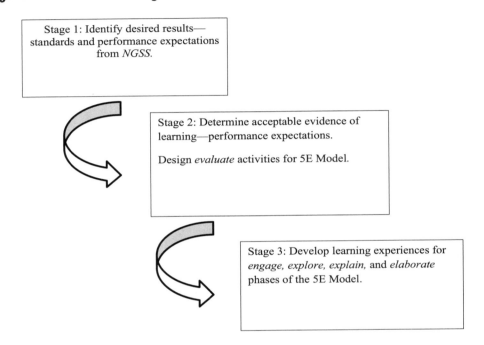

Stage 1: Identify desired results—standards and performance expectations from *NGSS*.

Stage 2: Determine acceptable evidence of learning—performance expectations.

Design *evaluate* activities for 5E Model.

Stage 3: Develop learning experiences for *engage, explore, explain,* and *elaborate* phases of the 5E Model.

Source: Adapted from Wiggins and McTighe 2005.

Design an Instructional Sequence

This process uses the BSCS 5E Instructional Model, the sequence of which is summarized in Figure 9.2. More elaborate descriptions are included in the following activities.

Figure 9.2. The 5E Instructional Model and an Instructional Sequence

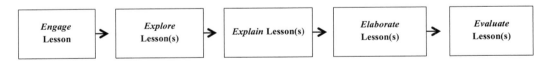

| *Engage* Lesson | *Explore* Lesson(s) | *Explain* Lesson(s) | *Elaborate* Lesson(s) | *Evaluate* Lesson(s) |

First, Identify a Series of Lessons You Already Teach and Design an Assessment.

You should identify a topic and lessons you already teach. Be sure you are comfortable with the lessons. What should students know and be able to do as a result of your teaching this series of lessons? What are the essential learning outcomes? What assessments are aligned with the lessons?

Ask yourself, "What would I accept as evidence indicating students achieved the learning outcomes of the lesson?" It is important to determine the means you could use to obtain evidence of learning. Traditionally, teachers have used test questions to identify what students have and have not learned. In designing the *evaluate* activity, you may use other experiences, such as a final project, web search, analysis of a video, or computer simulation. Use the means you think are most appropriate to obtain evidence of learning—for example, a response to questions, a written summary, a diagram, or a model. This is the *evaluate* phase, and most likely an addition to your original series of lessons.

Complete the information in Figure 9.3.

Figure 9.3. BSCS 5E Instructional Model and an Instructional Sequence: Evaluating the Concepts and Skills

| EVALUATION
A BRIEF DESCRIPTION	DETAILED DESCRIPTION OF INSTRUCTION
Students and teachers have opportunities to assess understandings and skills. How did the students—and you—do?	This phase emphasizes students assessing their understanding and abilities and provides opportunities for teachers to evaluate students' understanding of concepts and development of goals identified in learning outcomes. Be sure to clearly identify evidence for learning and the most effective means to collect that evidence.

Describe any modifications or additions to the evaluation used as the basis for the instructional sequence.

Second, Identify an Experience That Will Create a Teachable Moment.

Next, turn on your education imagination and find a question, demonstration, picture, simulation, or activity that you think will create a teachable moment for the students related to your learning goals. Got it? Okay, you have the *engage* phase of the instructional sequence based on the original lesson.

Complete the information in Figure 9.4 (p. 106).

Figure 9.4. BSCS 5E Instructional Model and an Instructional Sequence: Engaging the Learner

ENGAGE A BRIEF DESCRIPTION	DETAILED DESCRIPTION OF INSTRUCTION
Interest in a concept or problem is generated and students' current understanding is assessed. How can you create a teachable moment for the students?	This lesson initiates the learning task. The activities should (1) activate prior knowledge and make connections between past and present learning experiences, and (2) anticipate activities and focus students' thinking on the learning outcomes of current activities. The learners should become mentally engaged in the concepts, practices, abilities, and skills of the instructional unit.

An *Engaging* Lesson

Third, Modify a Lesson So Students Have an Opportunity to Explore.

Think of an activity where students will use their ideas and skills and try to solve the problem, answer the question, respond to the engaging experience, and further the learning goals. Then, provide opportunities to achieve the evaluation designed in step 2. Remember that students do not have to get the right solution or answer. In fact, you may identify various conceptions and less-than-adequate solutions. This is the point of the *explore* phase.

Complete the information in Figure 9.5.

Figure 9.5. BSCS 5E Instructional Model and an Instructional Sequence: Exploring the Concepts and Skills

EXPLORATION A BRIEF DESCRIPTION	DETAILED DESCRIPTION OF INSTRUCTION
Students participate in activities to explore questions or problems related to the learning outcomes. Are students motivated and expressing their interest and ideas?	This phase provides students with a common base of experiences within which they identify and begin developing concepts, practices, abilities, and skills. Students actively explore the contextual situation through investigations, reading, web searches, and discourse with peers.

Explore Lesson(s)

Fourth, Develop an Explanation.

This activity begins with the students giving their proposed solutions or answers to the *explore* activity. Then, you provide an *explanation*. This explanation should be clear and well illustrated, make connections to students' prior experiences using accurate terminology, and include the important concepts and abilities. Again, your explanation should be simple and direct.

Complete the information in Figure 9.6 (see p. 108).

Figure 9.6. BSCS 5E Instructional Model and an Instructional Sequence: Explaining the Concepts and Skills

EXPLANATION A BRIEF DESCRIPTION	DETAILED DESCRIPTION OF INSTRUCTION
Students express their understanding and develop evidence-based explanations. The teachers also provide definitions and explanations. Go ahead and clarify the concepts and practices, but make it clear.	This phase focuses on developing an explanation for the activities and situations students have been exploring. They verbalize their understanding of the concepts and practices. The teacher introduces formal labels, definitions, and explanations for concepts, practices, skills, and abilities.

Explanation Lesson(s)

Fifth, Apply the Explanations to New Situations.

Finally, provide a new but related activity where students apply their knowledge and skills. This is the *elaborate* phase. Again, make sure this activity promotes the learning goals and helps students successfully complete the evaluation task.

Complete the information in Figure 9.7.

Figure 9.7. BSCS 5E Instructional Model and an Instructional Sequence: Elaborating the Concepts and Skills

ELABORATION A BRIEF DESCRIPTION	DETAILED DESCRIPTION OF INSTRUCTION
Students have opportunities to expand and apply their understanding of the concepts within new contexts and situations. Give students a new challenge.	These lessons extend students' conceptual understanding through opportunities to apply knowledge, skills, and abilities. Through new experiences, the learners transfer what they have learned and develop broader and deeper understanding of concepts about the contextual situation and refine their skills and abilities.

Elaboration Lesson(s)

You began with lessons you currently teach and added activities and experiences based on criteria for the 5E phases. Now, try the lesson sequence with your students and make appropriate changes based on students' achievement of the learning goals.

Evaluating the Instructional Sequence

After implementing the revised lessons in your class, complete the summary of the instructional sequence (see Figure 9.8). Then answer the following questions:

- Did students have adequate and appropriate time and opportunity to develop an understanding of
  - › the basic concepts?
  - › the skills and abilities?
- Was instruction aligned with the assessments (*evaluate* phase)? Were the students successful in completing the evaluation as planned?
- Were there opportunities to make connections to *Common Core State Standards* for English language arts or mathematics?

Figure 9.8. A Summary and Analysis of Lessons

PHASES OF 5E INSTRUCTIONAL MODEL	WHAT DID THE TEACHER DO?	WHAT DID THE STUDENTS DO?	WHAT WAS THE PURPOSE OF THIS PHASE?	HOW WOULD YOU IMPROVE THIS ACTIVITY?
ENGAGE				
EXPLORE				
EXPLAIN				
ELABORATE				
EVALUATE				

Evaluating Your Understanding and Use of the 5E Model

The purpose here is not to evaluate the lesson(s). Think about the experience of designing an instructional sequence based on the BSCS 5E Instructional Model. Answer the following questions:

- What was the easiest part of designing the instructional sequence?
- What was the most difficult issue you encountered in this design process?
- What did you learn when you taught lessons based on the instructional model?
- If you had to design other units of instruction, what would you do differently?
- How would you describe your understanding of the BSCS 5E Instructional Model?
- How would you describe your use of the BSCS 5E Instructional Model?

CONCLUSION

The ultimate evaluation for this book is your understanding and use of the BSCS 5E Instructional Model. This chapter centered on evaluations that included adapting current lessons to accommodate the purpose of each phase of the 5E Model and your evaluation of that instructional sequence. Most important, the chapter recommended implementation of the instructional sequence in your classroom.

In the end, the chapter represents a self-evaluation of your understanding and use of the 5E Model. If you implemented and evaluated the 5E Model in your classroom, I congratulate you. Your initiative and courage represent the best of teaching.

REFERENCE

Wiggins, G., and J. McTighe. 2005. *Understanding by design.* Alexandria, VA: Association for Supervision and Curriculum Development (ASCD).

Conclusion

Teachers in all disciplines, grades, and schools hold the aim of improving student learning. This is, after all, why they teach. As a means to this goal, teachers also strive to find new lessons, activities, and curriculum materials that will enhance student learning. As part of these continued efforts to become better teachers, the use of coordinated and coherent lessons—learning cycles, integrated instructional sequences, and instructional models—has gained popularity and use.

Recent research reports, such as *How People Learn: Brain, Mind, Experience, and School* (NRC 2005), have confirmed what educators have asserted for many years: The sustained use of an effective, research-based instructional model can help students learn fundamental concepts in science and other disciplines. If we accept that premise, then an instructional model must be effective, supported with relevant research, *and* implemented consistently and widely to have the desired effect on teaching and learning.

Since the late 1980s, the Biological Sciences Curriculum Study (BSCS) has used one instructional model extensively in the development of new curriculum materials and professional development experiences. That model is commonly referred to as the BSCS 5E Instruction Model and consists of five phases: *engage, explore, explain, elaborate,* and *evaluate*. Each phase has a specific function and contributes to the teacher's coherent instruction and to the learners' formulation of a better understanding of scientific and technological knowledge, attitudes, and skills. The model frames a sequence and organization for programs, units, and lessons. Once internalized, it also can inform the many instantaneous decisions that teachers must make in classroom situations.

The BSCS 5E Instructional Model is grounded in sound education theory, has a strong and growing base of research to support its effectiveness, and has had a significant impact on education. Although encouraging, these conclusions indicate the need to continue conducting research on the effectiveness of the model, including when and how it is used and its effectiveness with different subjects, groups, and grades. There is a need to continue refining the model based on research and adapting the basic model in light of contemporary innovations such as the *Next Generation Science Standards* (*NGSS*), *Common Core State Standards* (*CCSS*), 21st-century skills, and STEM education.

The uniqueness of the BSCS 5E Instructional Model is, in part, related to its alliterative nature. Every stage of the model begins with the same letter—*E*. When compared to Herbart's (1901) model of preparation, presentation, generalization, and application or Atkin and Karplus's (1962) model of exploration, invention, and discovery, it becomes clear that the 5E Model has the potential to continue influencing classroom teaching. A danger, of course, is that something that is catchy and easy to remember might be misused as often as it is used effectively; however, something that cannot be remembered or understood is less likely to have widespread sustainable effects.

I also believe that the extensive use of the BSCS model can be explained by the fact that issues of teaching have significant personal meaning to all teachers; thus, there is an interest and motivation to embrace an instructional model that is immediately understandable and usable.

The five phases of the BSCS 5E Instructional Model are designed to facilitate the process of learning. The use of this model brings coherence to different teaching strategies, provides connections among educational activities, and helps teachers make decisions about interactions with students. The 5E Model allows ample time and provides multiple opportunities for learning to take place.

Studies of the 5E Model showed positive trends for student mastery of subject matter and interest in science. The most significant finding, however, is that there is a relationship between fidelity of use and student achievement. In other words, the BSCS 5E Instructional Model is more effective for improving student achievement when the teacher uses curriculum materials and the instructional sequence the way they were designed and developed. Without fidelity of use, the potential results of the model are diminished. This is a line of research that should be pursued. In addition, the research base for the BSCS 5E Instructional Model should be elaborated on through additional studies that compare its effect on mastery of subject matter, scientific reasoning, and interest and attitudes with other modes of instruction. The widespread use of the BSCS 5E Instructional Model warrants a commitment to a line of research that rivals that of other major initiatives in American education.

One noticeable void in the literature is research exploring the utility of the BSCS 5E approach in helping students develop an understanding of the nature of science and the complexity and ambiguity of empirical work, as well as 21st-century abilities such as teamwork skills, problem solving, and systems thinking. I also recommend exploratory studies on its effectiveness relative to the *CCSS*.

The range of applications of the BSCS 5E Instructional Model is one way to gauge the impact of the model. Across the decades the impact has been significant; to say the least, it has been applied in diverse contexts to a variety of disciplines. In addition, it serves as an indicator of its success and an instructional model in education. The BSCS 5E Instructional Model has become the foundation for a vast number of curriculum materials, state frameworks, and teacher-developed lessons and units used in education and, consequently, has had a large impact on the teaching and learning of science throughout the United States and internationally.

The early development of the BSCS 5E Instructional Model had the aim of designing an instructional sequence that would help teachers approach instruction in a meaningful way, one that enhanced student learning. I still hold this goal. At the time of its origin, I had no idea of the model's potential for widespread use by educators. Many within the science education community have recognized the model's practical value and incorporated it into school programs, state frameworks, and national guidelines. There is something to the model that has held the communities' interest during the decades, and this has touched me deeply. I encourage the continued use of the model with the full recognition that classroom teachers will bring appropriate adaptations based on the unique circumstances of their students. Classroom teachers can use the BSCS 5E Instructional Model to help them create teachable moments and, thus, enhance learning for all students.

Teachers have the challenge of teaching more effectively. They also must be accountable. Implementing the BSCS 5E Instructional Model is one way to improve your teaching. You can take a small step and begin using the 5E Model. You can enhance student learning and take control of accountability. You can be in a position to not only create teachable moments but also extend those moments with more time and significant opportunities to learn. In the end, if students are learning, accountability should not be an issue.

Effective teaching is an aim that all teachers consider. Implementing the BSCS 5E Instructional Model is one means to achieve this professional goal.

REFERENCES

Atkin, J. M., and R. Karplus. 1962. Discovery of invention? *The Science Teacher* 29 (5): 45.

Herbart, J. 1901. *Outlines of educational doctrine.* Edited by A. Lange. Translated by C. DeGarmo. New York: Macmillan.

National Research Council (NRC). 2005. *How people learn: Brain, mind, experience, and school.* Washington, DC: National Academies Press.

Sample 5E Model: Earth's Heat Engine

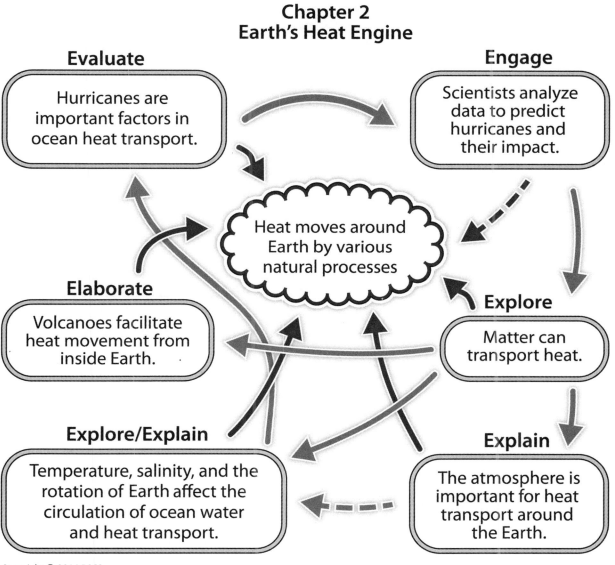

Chapter 2
Earth's Heat Engine

Evaluate
Hurricanes are important factors in ocean heat transport.

Engage
Scientists analyze data to predict hurricanes and their impact.

Heat moves around Earth by various natural processes

Elaborate
Volcanoes facilitate heat movement from inside Earth.

Explore
Matter can transport heat.

Explore/Explain
Temperature, salinity, and the rotation of Earth affect the circulation of ocean water and heat transport.

Explain
The atmosphere is important for heat transport around the Earth.

Copyright © 2014 BSCS.

APPENDIX 2

Sample 5E Model: Star Power

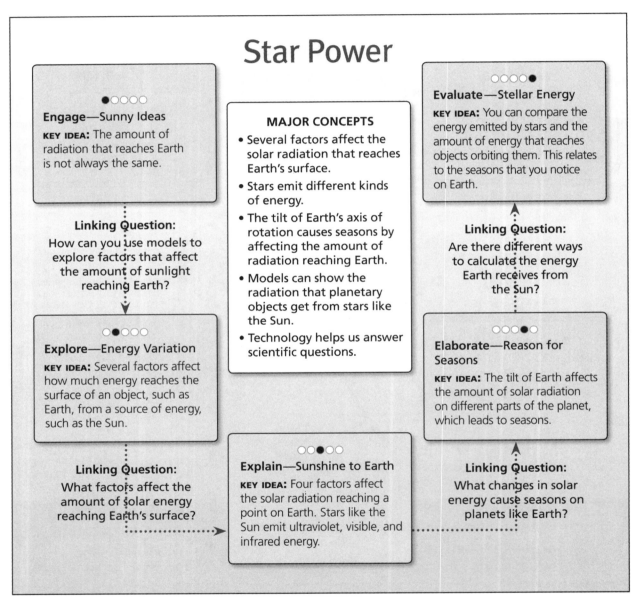

Star Power

Engage—Sunny Ideas

KEY IDEA: The amount of radiation that reaches Earth is not always the same.

Linking Question:
How can you use models to explore factors that affect the amount of sunlight reaching Earth?

Explore—Energy Variation

KEY IDEA: Several factors affect how much energy reaches the surface of an object, such as Earth, from a source of energy, such as the Sun.

Linking Question:
What factors affect the amount of solar energy reaching Earth's surface?

MAJOR CONCEPTS
- Several factors affect the solar radiation that reaches Earth's surface.
- Stars emit different kinds of energy.
- The tilt of Earth's axis of rotation causes seasons by affecting the amount of radiation reaching Earth.
- Models can show the radiation that planetary objects get from stars like the Sun.
- Technology helps us answer scientific questions.

Explain—Sunshine to Earth

KEY IDEA: Four factors affect the solar radiation reaching a point on Earth. Stars like the Sun emit ultraviolet, visible, and infrared energy.

Evaluate—Stellar Energy

KEY IDEA: You can compare the energy emitted by stars and the amount of energy that reaches objects orbiting them. This relates to the seasons that you notice on Earth.

Linking Question:
Are there different ways to calculate the energy Earth receives from the Sun?

Elaborate—Reason for Seasons

KEY IDEA: The tilt of Earth affects the amount of solar radiation on different parts of the planet, which leads to seasons.

Linking Question:
What changes in solar energy cause seasons on planets like Earth?

Sample 5E Model: Energy for You

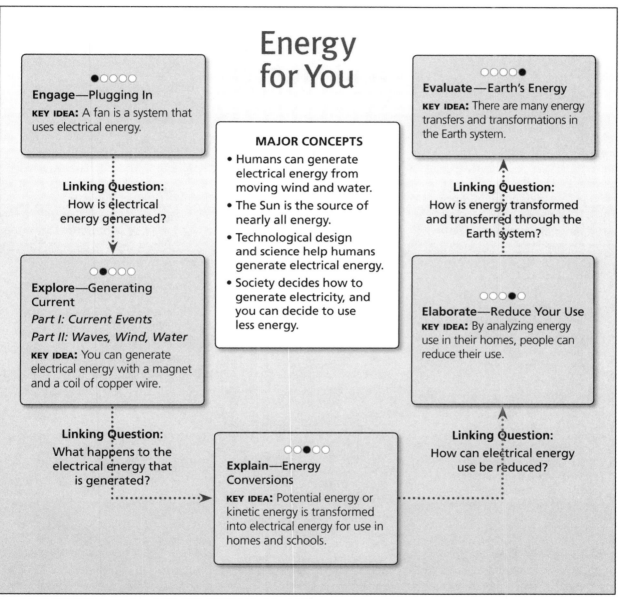

Energy for You

Engage—Plugging In

KEY IDEA: A fan is a system that uses electrical energy.

Linking Question:
How is electrical energy generated?

Explore—Generating Current

Part I: Current Events

Part II: Waves, Wind, Water

KEY IDEA: You can generate electrical energy with a magnet and a coil of copper wire.

Linking Question:
What happens to the electrical energy that is generated?

MAJOR CONCEPTS

• Humans can generate electrical energy from moving wind and water.

• The Sun is the source of nearly all energy.

• Technological design and science help humans generate electrical energy.

• Society decides how to generate electricity, and you can decide to use less energy.

Explain—Energy Conversions

KEY IDEA: Potential energy or kinetic energy is transformed into electrical energy for use in homes and schools.

Evaluate—Earth's Energy

KEY IDEA: There are many energy transfers and transformations in the Earth system.

Linking Question:
How is energy transformed and transferred through the Earth system?

Elaborate—Reduce Your Use

KEY IDEA: By analyzing energy use in their homes, people can reduce their use.

Linking Question:
How can electrical energy use be reduced?

APPENDIX 4

Sample 5E Model: Ecosystems and Energy

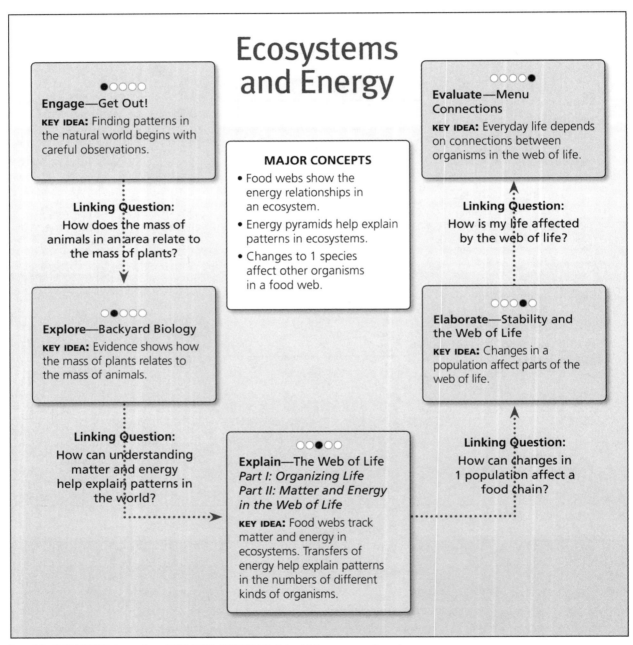

Ecosystems and Energy

Engage—Get Out!

KEY IDEA: Finding patterns in the natural world begins with careful observations.

Linking Question:
How does the mass of animals in an area relate to the mass of plants?

MAJOR CONCEPTS
- Food webs show the energy relationships in an ecosystem.
- Energy pyramids help explain patterns in ecosystems.
- Changes to 1 species affect other organisms in a food web.

Explore—Backyard Biology

KEY IDEA: Evidence shows how the mass of plants relates to the mass of animals.

Linking Question:
How can understanding matter and energy help explain patterns in the world?

Explain—The Web of Life
Part I: Organizing Life
Part II: Matter and Energy in the Web of Life

KEY IDEA: Food webs track matter and energy in ecosystems. Transfers of energy help explain patterns in the numbers of different kinds of organisms.

Evaluate—Menu Connections

KEY IDEA: Everyday life depends on connections between organisms in the web of life.

Linking Question:
How is my life affected by the web of life?

Elaborate—Stability and the Web of Life

KEY IDEA: Changes in a population affect parts of the web of life.

Linking Question:
How can changes in 1 population affect a food chain?

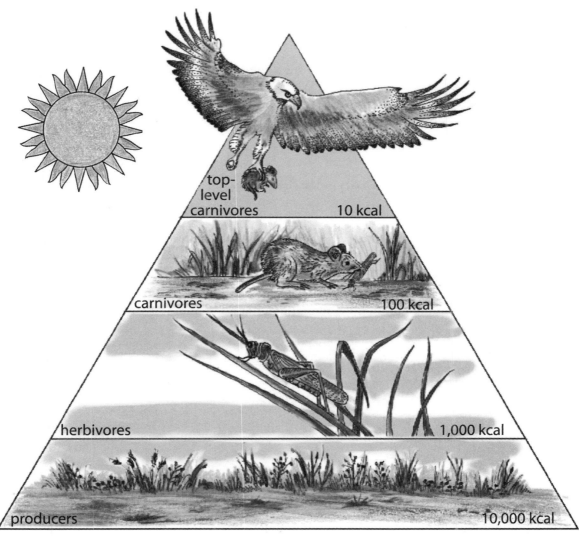

top-level carnivores — 10 kcal

carnivores — 100 kcal

herbivores — 1,000 kcal

producers — 10,000 kcal

APPENDIX 5

Sample 5E Model: Electrical Connections

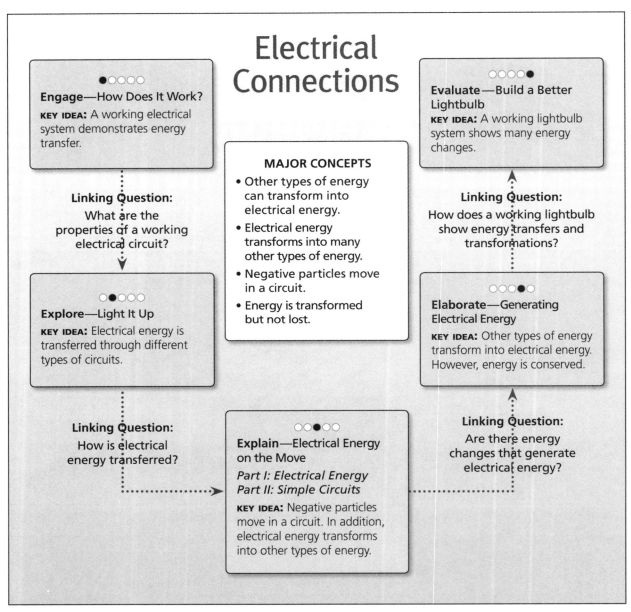

Electrical Connections

Engage—How Does It Work?
KEY IDEA: A working electrical system demonstrates energy transfer.

Linking Question:
What are the properties of a working electrical circuit?

Explore—Light It Up
KEY IDEA: Electrical energy is transferred through different types of circuits.

Linking Question:
How is electrical energy transferred?

MAJOR CONCEPTS
• Other types of energy can transform into electrical energy.
• Electrical energy transforms into many other types of energy.
• Negative particles move in a circuit.
• Energy is transformed but not lost.

Explain—Electrical Energy on the Move
Part I: Electrical Energy
Part II: Simple Circuits
KEY IDEA: Negative particles move in a circuit. In addition, electrical energy transforms into other types of energy.

Evaluate—Build a Better Lightbulb
KEY IDEA: A working lightbulb system shows many energy changes.

Linking Question:
How does a working lightbulb show energy transfers and transformations?

Elaborate—Generating Electrical Energy
KEY IDEA: Other types of energy transform into electrical energy. However, energy is conserved.

Linking Question:
Are there energy changes that generate electrical energy?

INDEX

*Page numbers in **boldface** type refer to figures or tables.*